Chemical Engineering Handbook

Chemical Engineering Handbook

Ian Goodwin

Willford Press
www.willfordpress.com

Published by Willford Press,
118-35 Queens Blvd., Suite 400,
Forest Hills, NY 11375, USA

ISBN: 978-1-64728-329-2

Cataloging-in-Publication Data

Chemical engineering handbook / Ian Goodwin.
p. cm.
Includes bibliographical references and index.
ISBN 978-1-64728-329-2
1. Chemical engineering. 2. Chemistry, Technical. 3. Metallurgy. 4. Engineering. I. Goodwin, Ian.
TP155 .C44 2022
660--dc23

For information on all Willford Press publications visit our website at www.willfordpress.com

Willford Press

Contents

Preface

The branch of engineering which deals with producing, using, designing, transforming and transporting energy and materials is known as chemical engineering. It utilizes principles of chemistry, physics, biology, economics and mathematics. Some of the key concepts within this field are process design and analysis, transport phenomena, and plant design and construction. Process design deals with defining the type and size of equipment, and the materials of construction. Plant design and construction involves creating plans, specifying and providing economic analyses for plant modifications, pilot plants and new plants. Chemical engineering is primarily applied in the transformation of raw materials into usable products. This book covers in detail some existent theories and innovative concepts revolving around chemical engineering. It presents this complex subject in the most comprehensible and easy to understand language. Those in search of information to further their knowledge will be greatly assisted by this book.

A detailed account of the significant topics covered in this book is provided below:

Chapter 1- The branch of engineering which seeks to produce, design and transform energy and materials by applying principles from chemistry, physics, mathematics and biology is termed as chemical engineering. This chapter has been carefully written to provide an easy introduction to the varied facets of chemical engineering.

Chapter 2- There are three major sub-disciplines within chemical engineering. These are biochemical engineering, electrochemical engineering and process engineering. This chapter closely examines the key concepts of these sub-disciplines of chemical engineering to provide an extensive understanding of the subject.

Chapter 3- The specialty within chemical engineering which focuses on studying and optimizing chemical reactions for the purpose of defining the best reactor design is known as chemical reaction engineering. All the diverse areas of study related to chemical reaction engineering such as chemical reactions, chemical kinetics and chemical reactors have been carefully analyzed in this chapter.

Chapter 4- The study of the flow of heat in diverse process situations is known as heat transfer. A few ways in which transfer of heat takes place are conduction, convection and radiation. Mass transfer refers to the net movement of mass from one location to another. This chapter discusses in detail these ways of heat transfer as well as the coefficients of mass transfer.

Chapter 5- Chemical engineering is applied in numerous fields such as agriculture and food industry. Within agriculture, it is used to produce agricultural chemicals like insecticides, fungicides and pesticides. The diverse applications of chemical engineering in these sectors have been thoroughly discussed in this chapter.

It gives me an immense pleasure to thank our entire team for their efforts. Finally in the end, I would like to thank my family and colleagues who have been a great source of inspiration and support.

Ian Goodwin

1

Chemical Engineering: An Introduction

The branch of engineering which seeks to produce, design and transform energy and materials by applying principles from chemistry, physics, mathematics and biology is termed as chemical engineering. This chapter has been carefully written to provide an easy introduction to the varied facets of chemical engineering.

Chemical engineering is a multi-disciplinary branch of engineering that combines natural and experimental sciences (such as chemistry and physics), along with life sciences (such as biology, microbiology and biochemistry) plus mathematics and economics to design, develop, produce, transform, transport, operate and manage the industrial processes that turn raw materials into valuable products. Many of the processes within chemical engineering involve chemical reactions, and the field takes cues from chemists who are looking for new ways to create products and to investigate the mechanisms within chemical reactions.

They are involved in a wide range of industries that includes:

- Pharmaceuticals
- Alternative Fuels
- Healthcare
- Manufacturing
- Polymers
- Environmental Health and Safety
- Biotechnology
- Food Processing
- Advanced Materials

Chemical engineering transforms raw materials into useful, innovative products for public and commercial consumption.

The fundamental principles of chemical engineering underlie the operation of processes extending well beyond the boundaries of the chemical industry, and chemical engineers are employed in a range of operations outside traditional areas. Plastics, polymers, and synthetic fibres involve chemical-reaction engineering problems in their manufacture, with fluid flow and heat transfer considerations dominating their fabrication. The dyeing of a fibre is a mass-transfer problem. Pulp and paper manufacture involve considerations of fluid flow and heat transfer. While the scale and materials are different, these again are found in modern continuous production of foodstuffs. The pharmaceuticals industry presents chemical engineering problems, the solutions of which have been essential to the availability of modern drugs. The nuclear industry makes similar demands on the chemical engineer, particularly for fuel manufacture and reprocessing. Chemical engineers are involved in many sectors of the metals processing industry, which extends from steel manufacture to separation of rare metals.

Further applications of chemical engineering are found in the fuel industries. In the second half of the 20th century, considerable numbers of chemical engineers have been involved in space exploration, from the design of fuel cells to the manufacture of propellants. Looking to the future, it is probable that chemical engineering will provide the solution to at least two of the world's major problems: supply of adequate fresh water in all regions through desalination of seawater and environmental control through prevention of pollution.

Chemical Process

Processes are the main concern of chemical engineers. A process is a system that converts feedstocks of lower intrinsic value to products of higher value. For instance, the block diagram of a process to manufacture ammonia from natural gas is shown in figure. This process has several sections, each one of which carries out a specific task and is, in effect, a mini-process. Natural gas, steam and air are fed to the Reformer Section that converts these feeds into a mixture of H_2, CO, CO_2, N_2, and H_2O. The overall chemical reactions involved are:

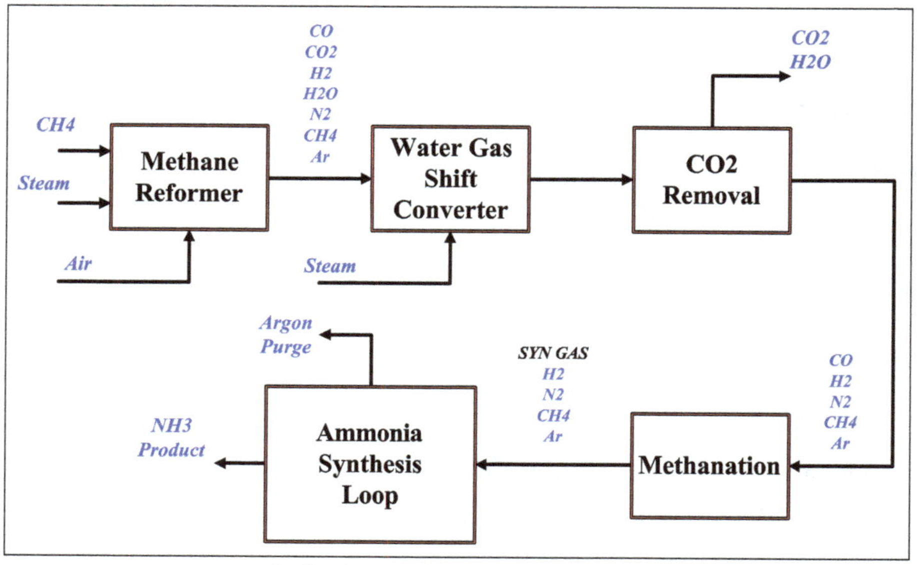

Block Diagram for Ammonia Process.

$$CH_4 + H_2O \rightarrow CO + 3H_2$$

$$CH_4 + O_2 \rightarrow CO + 2H_2$$

$$CH_4 + 2O_2 \rightarrow CO_2 + 2H_2O.$$

Since H_2 is the desired raw material from which to make ammonia, this gas mixture is sent to the CO Shift Section where additional steam is added to improve conversion by the water gas shift reaction:

$$CO + H_2O \rightarrow H_2 + CO_2.$$

The CO_2 and H_2O present in the gas mixture leaving the CO Shift Section are removed in the CO_2 Removal Section. The gas mixture leaving the CO_2 Removal Section contains primarily a 3/1 mixture of H_2 and N_2 (the N_2 coming from the air fed to the Reformer Section). It also contains small amounts of CO (the water gas shift reaction is equilibrium limited) as well as argon from the air feed to the Reformer Section.

The CO must be removed from this mixture because it will deactivate the catalyst used in the ammonia converter. This is done in the Methanation Section via the reaction:

$$CO + 3H_2 \rightarrow CH_4 + H_20.$$

The gas mixture leaving the Methanation Section contains a 3/1 mixture of H_2 and N_2 and trace amounts of CH_4 and Ar. This is sent to the NH_3 Synthesis Loop where the ammonia is made via the well-known reaction:

$$N_2 + 3H_2 \rightarrow 2NH_3.$$

So far each section of the process has been considered as a black box.

The ammonia synthesis reaction is equilibrium limited to the point where it must be run at rather high pressures (3000 PSIA or 20 mPa) in order to achieve a reasonable conversion of Syn Gas to ammonia across the Converter. Since the front end of the process is best run at much lower pressures, a Feed Compressor is needed to compress the Syn Gas to the operating pressure of the Synthesis Loop.

Due to the unfavorable reaction equilibrium, only part of the Syn Gas is converted to ammonia on a single pass through the Converter. Since the unconverted Syn Gas is valuable, the majority of it is recycled back to the Converter. Due to pressure drop through the Synthesis Loop equipment, a Recycle Compressor is required to make up this pressure drop. The recycled Syn Gas is mixed with fresh Syn Gas to provide the feed to the Synthesis Converter.

The effluent from the Synthesis Converter contains product ammonia as well as unreacted Syn Gas. These must be separated. At the pressure of the synthesis loop ammonia can be condensed at reasonable temperatures. This is done in the Ammonia Condenser where the Syn Gas is cooled to approximately ambient temperature using cooling water (CW). The liquid ammonia is then separated from the cycle gas in the Ammonia Knockout (KO) Drum. The liquid ammonia is removed from the bottom of the drum and the cycle gas leaves the top.

Some of the cycle gas must be purged from the Synthesis Loop. Otherwise, the argon that enters the loop in the Syn Gas has no way to leave and will build up in concentration. This will reduce the rate of the ammonia synthesis reaction to an unacceptable level. To prevent this from happening, a small amount of the cycle gas must be purged, the amount being determined by the amount of argon in the feed and its acceptable level in the Synthesis Converter feed (generally about 10 mol %).

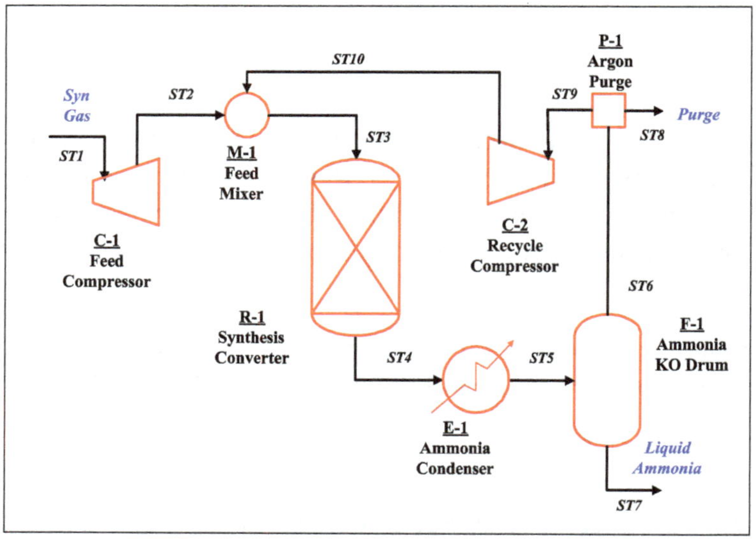

Ammonia synthesis loop process flow diagram.

This description of the Ammonia Synthesis Loop covers only the most important aspects. Modern ammonia plants are much more complex due to attention paid to maximizing the amount of ammonia produced per mol of Syn Gas fed to the loop. However, the process shown in figure is typical of many chemical processes.

Basic Processing Functions

Several basic processing activities are required by the Ammonia Synthesis Loop in order to convert the hydrogen and nitrogen to ammonia product. These activities are common to almost all chemical processes; the functionality of each can be considered independently of any specific process.

There are five of these processing activities that are of major interest in chemical engineering:

1. Chemical Reaction
2. Mixing
3. Separation
4. Materials Transfer (Fluid flow)
5. Energy (Heat) Transfer

Of these the first three are involved in the process material balance. The last two are necessary adjuncts to operation of chemical processes. Material must be transported from one piece of equipment to another. For the large number of chemical plants processing only liquids and vapors, this involves fluid flow. Also, streams must be heated to or cooled to specified temperatures as dictated by the needs of the process. For example, reactors in general are operated at temperatures higher than those that are acceptable for most separation operations. Thus, streams must be heated to reaction temperature and then cooled back down for subsequent processing.

Chemical Reaction

What distinguishes the chemical process industries from almost all others is the use of chemical reactions to convert less valuable raw materials to more valuable products. In other words, chemical reaction is the heart and soul of almost all processes.

In the Ammonia Synthesis Loop example, R-1, the Synthesis Converter is where the ammonia synthesis reaction takes place.

Mixing

Many chemical reactions involve two or more reactants. In order for these reactants to react, these must be brought into contact at the molecular level, i.e., mixed, before the desired reactions can proceed properly.

Mixing is also required if several substances are to be blended to create a product mixture with the desired properties.

Example process, mixing of fresh synthesis gas and recycled synthesis gas takes place in the Feed Mixer, M-1, before being sent to the Synthesis Converter.

Separation

In an ideal chemical process, exactly the right amounts of reactants would be mixed and reacted completely to the desired product. Unfortunately, this is seldom the case. Many reactions cannot be carried to completion for various reasons. Seldom do the reactants react only to the desired product. Unwanted byproducts are formed in addition to the target product. Finally, the reactants are seldom 100% pure, again for many reasons.

The result is that the material leaving the reactor is a mixture containing the desired product, by-products, unreacted raw materials, and impurities. It may also contain other components deliberately introduced for one reason or another. The use of a homogeneous catalyst is just one example of this.

This mixture must be separated into its various constituents. The product must be separated from almost everything else and brought to an acceptable level of purity such that it can be sold. The reactants, being valuable, must be recovered and recycled back to the reactor. The impurities and by-products must be separated out for disposal in a suitable manner.

Thus, the activity of separation is fundamental to the operation of almost any process. Some separation systems are relatively simple. Others constitute the major part of the process. As will be seen, separation takes on many forms.

Separation of ammonia from cycle gas takes place in the Ammonia KO Drum, F-1. The cycle gas leaving the Ammonia Condenser, E-1, contains droplets of liquid ammonia. This mixture enters the middle of the KO Drum. The vapor, being less dense, flows upward while the liquid ammonia falls to the bottom of the drum.

Materials Transfer

The Ammonia Synthesis Loop consists of several items of equipment, each of which has material flowing in and material flowing out. These flows take places though process piping connecting the various items of equipment. If the pressure in the upstream item of equipment is sufficiently higher than that in the downstream item, then material will flow from the upstream equipment to that downstream without the need for any additional equipment. This pressure difference is necessary to overcome the friction due to fluid flow through the piping. As a result, the pressure will decrease in the direction of flow through the process.

Thus, as synthesis gas flows from the inlet to the reactor (Stream 3) through the reactor, condenser and knockout drum, the pressure decreases significantly. In order to be able to recycle the unreacted hydrogen and nitrogen back to the reactor, some means is required to increase the pressure of this stream back to that of the reactor inlet. Hence the presence of the compressor in the recycle loops.

Unit Operations

One of the major contributions to the practice of chemical engineering is the concept of the unit operations. This concept was developed by Arthur D. Little and Warren K. Lewis in the early 1900's. Prior to its development, chemical engineering was, to a large extent, practiced along the lines of specific process technologies. For instance, if distillation was required in the manufacture of acetic acid, it became a problem in acetic acid distillation. The fact that a similar distillation might be required for the manufacture of, say, acetaldehyde was largely ignored. What Little and Lewis did was to show that the principles of distillation (as well as many other processing operations) were the same regardless of the materials being processed. So, if one knew how to design distillation columns, one could do so for acetic acid, acetaldehyde, or any other mixture of reasonable volatility with equal facility. The same proved to be true of other operations such as heat transfer by two-fluid heat exchangers, gas compression, liquid pumping, gas absorption, liquid-liquid extraction, fluid mixing, and many other operations common to the chemical industry.

The case for chemical reactors is less clear. Each reaction system tends to be somewhat unique in terms of its reaction conditions (temperature, pressure, type of catalyst, feed composition, residence time, heat effects, and equilibrium limitations). Thus, each reaction system must be approached on its own merits with regard to the choice of reactor type and design. However, the field of chemical reaction engineering has undergone substantial development over the past few decades, the result being that much of what is required for the choice and design of reactors is subject to a rational and quantitative approach.

Modes of Process Operation

There are several modes of process operation. The one that has been most widely studied by chemical engineers is that of the steady-state operation of continuous processes. The reasons have also been discussed. In this mode of operation we assume that the process is subject to such good control that as feed materials flow into the process as constant flow rates, the necessary reactions, separations, and other operations all take place at conditions that do not vary with time. The amount of material in each item of equipment (its inventory) does not vary with time; pressures and liquid levels are constant. Nor do temperatures, compositions, and flow rates at each point in the process vary with time. From the standpoint of the casual observer, nothing is happening. However, finished product is flowing out of the other end of the process into the product storage tanks.

The next most common mode of process operation is known as batch operation. Here, every processing operation is carried out in a discrete step. Reactants are pumped into a reactor, mixed, and heated up to reaction temperature. After a suitable length of time, the reactor turned off by cooling it down. It now contains a mixture of products, by-products, and unreacted reactants. These are pumped out of the reactor to the first of the various separation steps, possibly a batch distillation or a filtration if one of the products or byproducts is a solid. Each batch step has a beginning, time duration, and an end. (Most activities around the home are batch in nature - cooking, washing clothes, etc.).

A third mode is cyclic operation. From the standpoint of flows in and out of both the process and individual items of equipment, operation is continuous. However, in one or more items of equipment, operating conditions vary in time in a cyclical manner. A typical example is reactor whose catalyst deactivates fairly rapidly with time due to, say, coke formation on the catalyst. To recover the catalyst activity, it must be regenerated by being taken out of reaction operation. The coke is removed either by stripping by blowing an inert gas over the catalyst or, in the more difficult cases, by burning the coke off with dilute oxygen in an inert gas carrier.

Two characteristics of cyclical operation become apparent. First off, if the process is to be operated continuously but it one or more items of equipment must be taken off line for regeneration of one sort or another, we must have at least two of such items available in parallel. One is on line while another is being regenerated. The second characteristic is the is that operation conditions in the cyclically operated equipment must vary with time. If catalyst activity decrease with time, then something must be done to maintain the productivity of the reactor. Usually this is achieved by raising the reactor temperature. Thus, a freshly regenerated reactor will start off at a relatively low temperature; the temperature will be raised during the cycle; and the reactor will be taken off line when no further benefit is to be obtained by raising the temperature any further.

Unit Operations

Every industrial chemical process is based on Unit Operations (physical treatment) and Unit Process (chemical treatment) to produce economically a desired product from specific raw materials. The raw materials are treated through physical steps to make it suitable for chemical reaction. So, knowledge of unit operations like 'Mixing and agitation of liquid' and' heat flow' is very much necessary. The subject Unit Operations is based on fundamental laws, physicochemical principles. Unit Operations gives idea about science related to specific physical operation; different equipments-its design, material of construction and operation; and calculation of various physical parameters (mass flow, heat flow, mass balance, power and force etc.). Examples of Unit Operations are listed in table.

Table: List of some unit operations.

Heat flow	Mixing
Drying	Absorption
Evaporation	Adsorption
Distillation	Condensation
Crystallization	Vaporization
Leaching	Separation
Extraction	Sedimentation
Filtration	Crushing

Following are some examples of physical processes:

- Sugar Manufacture: Sugar cane crushing → sugar extraction → thickening of syrup → evaporation of water → sugar crystallization → filtration → drying → screening → packing.
- Salt Manufacture: Brine transportation → evaporation → crystallization → drying → screening → conveying → packaging.
- Pharmaceutical Manufacture: Formulation of chemicals, mixing, granulation → drying of granules → screening → pressing tablet → packaging.

The equipment used in the chemical processes industries can be divided into two classes:

- Proprietary equipment, such as pumps, compressors, filters, centrifuges and dryers, is designed and manufactured by specialist firms.
- Non-proprietary equipment is designed as special, one-off, items for particular processes; for example, reactors, distillation columns and heat exchangers.

Momentum, Heat and Mass Transfer

In most of the unit operations encountered in chemical and petroleum industries, or more of the processes of momentum, heat and mass transfer involved.

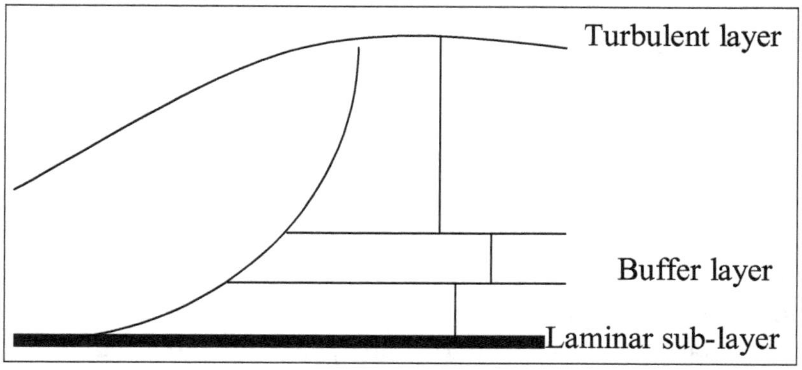

In some cases, momentum, heat and mass transfer all occur simultaneously as, for example:

In a water-cooling tower where transfer of sensible heat evaporation both take place from the surface of the water droplets. When a fluid flows under turbulent conditions over a surface, the flow can conveniently be divided into three regions:

- At the surface, the laminar sub-layer, in which the only motion at right angles to the surface is due to molecular diffusion.
- Next, the buffer layer, in which molecular diffusion and eddy motion are of comparable magnitude.
- Finally, over the greater part of the fluid the turbulent region in which eddy motion is large compared with molecular diffusion.

In addition to momentum, both heat and mass can be transferred either by molecular diffusion alone or by molecular diffusion combined with eddy diffusion. Because the effects of eddy diffusion are generally far greater than those of the molecular diffusion, the main resistance to transfer will lie in the region where only molecular diffusion is occurring.

Transfer by Molecular Diffusion Momentum Transfer

Momentum Transfer

When the flow characteristics of the fluid are Newtonian, the shear stress (Ry) in a fluid is proportional to the velocity gradient and to the viscosity:

$$R_y = -\mu \frac{dU_x}{dy}$$

$$R_y = -\frac{\mu}{\rho} \frac{d(\rho U_x)}{dy}$$

where,

U_x: is the velocity of the fluid parallel to the surface at distance (y) from it.

R_y: is the shear stress within the fluid, [is a measure of the rate of transfer of momentum per unit area at right angles at the surface.

The negative sign indicates that momentum is transferred from fast to the slow moving fluid and the shear stress acts in such a direction as to oppose the motion of the fluid.

Unit Process

Unit processes refers to processes that involve making chemical changes to materials, as a result of chemical reaction taking place. For instance, in the combustion of coal, the entering and leaving materials differ from each other chemically: coal and air enter, and flue gases and residues leave the combustion chamber. Combustion is therefore a unit process. Unit processes are also referred to as chemical conversions.

Together with unit operations (physical conversions), unit processes (chemical conversions) form the basic building blocks of a chemical manufacturing process. Most chemical processes consist of a combination of various unit operations and unit processes.

Material Balance

The material balance is the fundamental tool of chemical engineering. It is the basis for the analysis and design of chemical processes. So it goes without saying that chemical engineers must thoroughly master its use in the formulation and solution of chemical processing problems.

In chemical processing we deal with the transformation of raw materials of lower value into products of higher value and, in many, cases unwanted byproducts that must be disposed of. In addition many of these chemical compounds may be hazardous. The material balance is the chemical engineer's tool for keeping track of what is entering and leaving the process as well as what goes on internally. Without accurate material balances, it is impossible to design or operate a chemical plant safely and economically.

Conservation of Mass

The principle of conservation of mass is fundamental to all chemical engineering analysis. The basic idea is relatively easy to understand since it is fact of our everyday life.

Let us consider a simple example. Suppose we are required to prepare one kilogram of a solution of ethanol in water such that the solution will contain 40% ethanol by weight. So, we weigh out 400 grams of ethanol and 600 grams of water and mix the two together in a large beaker. If we weigh the resulting mixture (making appropriate allowance for the weight of the beaker), experience says it will weigh 1000 grams or one kilogram and it will. This is a manifestation of the conservation of mass.

"That, in the absence of nuclear reactions, mass is conserved is a fundamental law of nature."

Suppose we happened to measure the volumes involved in making up our alcohol solution. Assuming that we do this at 20 C, we would find that we added 598.9 ml of water to 315.7 ml of ethanol to obtain 935.2 ml of solution. However, the sum of the volumes of the pure components is 914.6 ml. We conclude that volume is not conserved.

If we were to separate it back into its pure components (something we could do, for instance, by azeotropic distillation) and did this with extreme care to avoid any inadvertent losses, we would obtain 400 gm of ethanol and 600 gm of water. Thus, in this case, not only was total mass conserved but the mass of each of the components was also.

This is not always true. Suppose that instead of adding ethanol and water, we added (carefully and slowly) sodium hydroxide to sulfuric acid. Suppose that the H_2SO_4 solution contains exactly 98.08 pounds of H_2SO_4 and that we add exactly 80.00 pounds of NaOH. A chemical reaction will take place as follows:

$$H_2SO_4 + 2NaOH \rightarrow Na_2SO_4 + 2H_2O.$$

Notice that the amount of H_2SO_4 in the original solution is 1.0 lb-mol and that the amount of NaOH added is exactly 2.0 lb-mols. What we are left with is 1.0 lb-mol of Na_2SO_4 or 142.05 lbs and 2.0 lb-mols of H_2O or 36.03 lbs. No individual component is conserved; the H_2SO_4 and the NaOH have disappeared and in their place we have Na_2SO_4 and H_2O. However, if we look at the atomic species H, O, S, and Na, we will find that these are all con-served. That is exactly what the reaction equation expresses.

Thus we have to be careful to identify the appropriate conserved species for the system we are analyzing. If no chemical reactions are involved, then each of the molecular species is conserved. If chemical reactions are involved, then only atomic species are conserved. There will be a mass balance for each of the conserved species. In the example above it does not make much difference since there are four conserved atomic species and four molecular species. But, if additional reactions take place involving, say, Na_2S and $NaHSO_4$, then the number of molecular species exceeds the number of conserved atomic species. This will generally be the case.

Control Volumes

We apply the principle of the conservation of mass to systems to determine changes in the state of the system that results from adding or removing mass from the system or from chemical reactions taking place within the system. The system will generally be the volume contained within a precisely defined section of a piece of equipment. We refer to this precisely defined volume as a control volume.

It may be the entire volume of the equipment. This would be the case if the system is a cylinder containing a gas or gas mixture. Or it may be the volume associated with a particular phase of the material held within the system. For instance, a flash drum is used to allow a mixture of vapor and liquid to separate into separate vapor and liquid phases. The liquid phase will occupy part of the total volume of the drum; the vapor, the remainder of the volume. If we are interested only in what happens to the liquid phase, then we would specify the volume occupied by the liquid as our control volume.

The control volume can change over the course of an operation. Suppose we are adding liquid to a tank that contains 100 Kg of water to start with and that we add another 50 Kg. The tank would originally contain 100 liters of water but would contain 150 liters after the addition. On the other hand, if our interest is in the entire contents of the tank - both the liquid and the vapor in the space above it - then we would take the volume of the tank itself as our control volume. This volume, of course, will not change.

Holdup or Inventory

Another concept that we will need to make precise is that of holdup, also known as inventory or accumulation. Holdup refers to the amount of a conserved species contained within a control volume. We can refer to the total holdup as simply the total mass of material contained within the control volume. Or we can refer to the holdup of a particular component, sodium chloride say, which is contained within the control volume. Needless to say, the sum of the holdups of all of the individual components within the control volume must equal the total holdup.

Material Balance Basis

Whenever we apply the principle of conservation of mass to define a material balance, we will want to specify the basis for it. Generally, the basis is either the quantity of total mass or the mass of a particular component or conserved species for which the material balances will be defined. Or, for continuous processes, it might be the mass flow rate of a component or conserved species.

Quite often the basis will be set by the specification of the problem to be solved. For instance, if we are told that a tank contains 5,000 pounds of a particular mixture about which certain questions are to be answered, then a natural basis for the problem would the 5,000 pounds of the mixture. Or, if we are looking at a continuous process to make 10,000 Kg/hr of ethanol, a reasonable choice for a basis would be this production rate.

Some problems, however, do not have a naturally defined basis so we must choose one. For instance, if we are asked what is the mass ratio of NaOH to H_2SO_4 required to produce a neutral solution of NaCl in water, we would have to specify a basis for doing the calculations. We might choose 98.08 Kg of H_2SO_4 (1.0 Kg-mol) as a basis or we could chose 1.0 lb of H_2SO_4. Either is acceptable. One basis may make the calculations simpler than another, but in this day of personal computers the choice is less critical than it might have been years ago. Whatever the choice of basis, it is mandatory that all material balances are defined to be consistent with it.

Material Balances

We are now in a position to define material balances for some simple systems. There are three basic situations for which we will want to do this:

1) A discrete process in which one or more steps are carried out over a finite but indefinite period of time. An example of such a process is the dissolving of a specified quantity of salt in a quantity of water contained in a tank. We are only interested in the concentration in weight % of the salt in the water after it is completely dissolved and not how long it takes for the salt to dissolve.

2) A continuous process operating in the steady state. By definition, continuous process operating in the steady state undergoes no changes in its internal state variables such as temperatures, pressures, compositions, and liquid levels. In addition, all the flow rates of all streams entering and leaving each item of equipment are constant. What this means from the standpoint of material balances is that there is no change in any of the holdups in the system.

3) Dynamic processes this is the general case in which both the holdups within equipment as well as the flow rates and compositions of the input and output streams can vary with time.

Examples:

Discrete Process

Let us now consider the application of the principle of the conservation of mass to a discrete process. For each of the conserved species, it can be stated as follows:

Change in holdup = additions to the control volume - withdrawals from the control volume

Consider the following example: We have a tank that initially holds 100 Kg of a solution containing 40% by weight of salt in water. We add 20 Kg of salt to the tank and allow it to dissolve. What do we now have in the tank?

First we have to identify the conserved species. Since there are no chemical reactions involved, both salt and water are conserved species. Next we have to define the control volume. It seems natural to choose the salt solution in the tank. Our basis is the amount of solution originally contained in the tank.

Now we can define a material balance for each conserved species as follows:

Water

Initial holdup of water = (1 - 0.4)(100) = 60 Kg.

Change in holdup of water = additions of water to the control volume - withdrawals of water from the control volume.

Since no water is either added or withdrawn, the change in this holdup is zero. Therefore the holdup of water after the salt addition is still 60 Kg.

Salt

Initial holdup of salt = (0.4)(100) = 40 Kg.

Change in holdup of salt = additions of salt to the control volume - withdrawals of salt from the control volume.

We add 20 Kg of salt and withdraw no salt. Therefore the change in the holdup of salt = +20 Kg and the holdup of salt after the addition is 40 + 20 = 60 Kg. A simple calculation shows that the concentration of salt in the tank is now 50 weight %.

We withdraw 40 Kg of the solution now in the tank. Since the solution in the tank is 50 weight % salt and 50 weight % water, in withdrawing 40 Kg of solution, we will withdraw 20 Kg of salt and 20 Kg of water. This will leave 60 - 20 = 40 Kg of each component in the tank. The composition has not changed from Step 1.

Continuous Steady-state Process

Let us consider a continuous mixer which has two input streams and, of course, one output stream. Suppose the first input stream has a flow rate of 10000 lb/hr of a 40 wt. % solution of salt in water while the second input stream has a flow rate of 20000 lb/hr of a 70 wt. % solution of salt in water. What is the flow rate and composition of the output stream?

Since the system is now characterized in terms of rates of flow into the control volume (additions) and rates of flow out of the system (withdrawals), we need to restate the principle of conservation of mass as follows:

Rate of change of holdup = rate of additions to the control volume - rate of withdrawals from the control volume.

For a continuous system operating in the steady state, the holdup does not change with time. Therefore, the rate of change of holdup is zero and equation above becomes:

Rate of withdrawals from the control volume = Rate of additions to the control volume.

Let us apply this to the mixer problem. The control volume is the contents of the mixer (even though these do change) and the basis is the total rate of flow to the mixer. As in the previous example, the conserved species are salt and water.

Salt

Rate of withdrawal of salt = rate of additions of salt to the mixer.

Rate of additions = (10000)(0.4) + (20000)(0.7) = 18000 lb/hr of salt.

Water

Rate of withdrawal of water = rate of additions of water to the mixer.

Rate of additions = (10000)(0.6) + (20000)(0.3) = 12000 lb/hr.

Thus the stream leaving the mixer has a flow rate of 18000 lb/hr of salt and 12000 lb/hr of water, for a total of 30000 lb/hr. This is exactly the total flow rate of the mixer output we get by adding up the total flow rates to the mixer. Also, the composition of salt of the stream leaving the mixer:

(100)(18000)/(30000) = 60 wt %

Dynamic Process

Consider the surge tank shown in figure. Water flows into the tank with a flow rate F_{in} lb/hr. It flows out at a rate F_{out} lb/hr. The flow rates in and out can be adjusted by means of the valves in the inlet and outlet piping. The tank has the form of an upright cylinder that has a cross section area of S ft^2 . The liquid level in the tank is z ft.

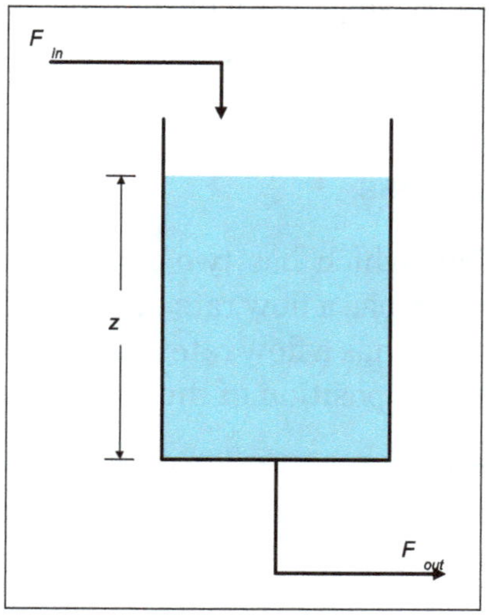

Surge Tank.

We know from experience that if the flow rates in and out are not exactly equal, the level in the tank will change with time. If the inlet flow rate exceeds the outlet flow rate, then the level will rise and vice versa. Now, the purpose of a surge tank is to absorb changes in the inlet flow rate while maintaining a relatively constant outlet flow rate.

(The reservoirs that supply water to a town or city are surge tanks where the inlet flow is the run-off from rainstorms and the outlet flow is the daily consumption by the town or city.) Thus, a question that designers of surge tanks must ask is, given an estimate of the variations of inlet and outlet flow rates as functions of time, how big must the surge tank be so that it never runs dry (town loses its water supply) or never overflows (area surrounding the reservoir is flooded). In general this is a complex design problem but let us look at a simple example to at least illustrate the concept.

Suppose that under normal conditions the level in the tank is to be half the height of the tank. If the flow rate into the tank becomes zero for a period of time (no rain), how long will it take for the tank to run dry if the outlet flow rate is maintained at its usual value? Specifically, suppose that the cross section area of the tank is 10 ft2 and its height is 10 ft and the normal outlet flow rate is 12,480 lb/hr.

First, let us take the volume of liquid in the tank as the control volume. The holdup of water in the control volume will be:

Holdup = Szρ, where ρ is the density of water (62.4 lb/ft^3).

Consider an interval of time Δt. Suppose that over that time interval the inlet and outlet flow rates are constant but not necessarily equal. Then, by the conservation of mass the change in the holdup will be given by:

$$\text{Holdup}|_{t=t} - \text{Holdup}|_{t=o} = F_{in}\,\Delta t - F_{out}\,\Delta t$$

Now, if we divide both sides of the mass balance equation ($\text{Holdup}|_{t=t} - \text{Holdup}|_{t=o} = F_{in}\,\Delta t - F_{out}\,\Delta t$) by Δt and take the limit as $\Delta t -> 0$, we get the differential form of the mass balance, to wit,

$$d[Sz\rho]/dt = F_{in} - F_{out}$$

If we assume that is constant (a reasonable assumption if the temperature is also reasonably constant), then our mass balance equation becomes:

$$dz/dt = (F_{in} - F_{out})/S\rho$$

For our problem $F_{in} = 0$ and $F_{out} = 12{,}480\text{ lb/hr}$, both constant. We can calculate dz/dt, that is:

dz/dt = (0 – 12480)/(10)(62.4) = –20 ft/hr.

Since the nominal level is 5.0 ft (half the tank height of 10ft), it will take 0.25 hour or 15 minutes for the tank to run dry.

Steady-state Process Modeling

Unit Models

Material balance calculations begin with the characterization of the individual unit operations by mathematical models. These are known as unit models or process blocks.

Consider the generalized process block shown in figure. Here, in1 and in2 are the numbers or names of the input streams to the process block and out1 and out2 are the numbers or names of the output streams. A block can have more or less than the two input streams shown; the same is true of output streams.

Also, the quantities P_1, P_2,..., P_m are a vector of parameters required to characterize the process block. For instance, if the block represents an isothermal flash, two parameters, namely the flash pressure and temperature, would have to be specified as part of the characterization of the block.

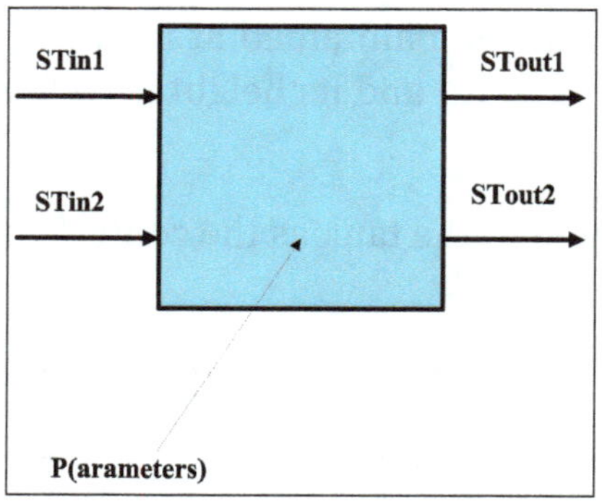

Generalized Unit Equipment Model.

A process unit operation block can now be characterized as follows:

Let X_n = the stream characterization vector for stream n.

Then

$$X_{out1} = G_{J,1}[X_{in1}, X_{in2}, ..., X_{inn}; P_1, P_2, \ldots, P_m]$$

$$X_{out2} = G_{J,2}[X_{in1}, X_{in2}, ..., X_{inn}; P_1, P_2, ..., P_m]$$

$$\vdots$$

$$X_{outn} = G_{J,n}[X_{in1}, X_{in2}, ..., X_{inn}; P_1, P_2, ..., P_m]$$

In general, these models are nonlinear and difficult to solve without using a computer.

Linear Input-Output Models

For many purposes such as preliminary material balance calculations for scoping out a design, simple linear unit models are perfectly adequate. These allow the material balance to be done by manual calculations (or by a spreadsheet program). This direct participation in the calculations at the early stages is recommended since in general the engineer will develop more insight into the workings of the process than if the calculations are done at one remove by a process simulation program.

Four simple models are all that are needed in order to do linear material balances. Actually only two are needed. The remaining two are special cases of the others.

Mixer (MIX)

In this model two or more streams are added together to produce a single output stream that is a mixture of all the input streams.

A mass balance on the ith component gives the equation describing the mixer, i.e.,

$$f_{out,i} = \sum_{j=1}^{nin} f_{in\ i,\ j}$$

Note that no parameters are required to characterize a mixer. This is not the case for any of the other equipment models.

Reactor (REACT)

A Reactor takes a feed stream, and by chemical reactions, converts some into other components. A mass balance on the ith component again gives the performance equation for the reactor, i.e.,

$$fi,_{out} = f_{i,in} + {}_{i}$$

where $_{i}$ = the extent of reaction of component i, i.e., the net number of mols of component i produced by reaction.

Note: $_{i}$ will be negative for reactants, positive for products.

Note also that REACT is a special case of MIX where i can be considered the flow rate of component i in the second input stream to MIX.

Or the $\grave{a}_i$ may be given as constant values based on a specified production rate and the reaction stoichiometry. If a mathematical model for the reaction kinetics is available and the reactor type has been chosen, the $\grave{a}_i$ can be estimated from a separate reactor calculation.

Separator (SEPAR)

A Separator is used to model process units in which each component in the feed is separated into two output streams.

The mass balance equations describing the performance of the Separator for the ith component are:

$$f_{i,out1} = s_i\ f_{i,in}$$

$$f_{i,out2} = (1 - s_i)\ f_{i,in}$$

where s_i = the separation coefficient for component i.

Note that conservation of mass dictates that 0 s_i 1.

As with the $_i$ for REACT, the s_i for SEPAR must be known or estimated by other means. For conceptual design material balances for which the separation equipment generally has not yet been selected, let alone designed, one will generally assume reasonable values, say 99% recovery of one key component in the overhead ($ST1_{out}$) and a similar recovery of the other key component in the bottoms (ST_{2out}).

Complex separators, such as those shown in figure, can be modeled by a combination of the models for MIX and SEPAR.

Flow Splitter (SPLIT)

In many processes, a stream is split into two smaller streams, each having the same composition as the input stream.

The governing equation for the Stream Splitter is an overall or total mass balance, i.e.,

$$F_{out1} = S\ F_{in}$$

The same relationship holds for each component. So, $f_{i,out1} = S\ f_{i,in}$ and

$$f_{i,out2} = (1 - S)\ f_{i,in}$$

Note that SPLIT is a special case of SEPAR for which s_i = S for all i. Also, 0 S 1. A flow splitter having more than two output streams can be written along similar lines.

The major difference between SEPAR and SPLIT is that the s_i are dictated by physical considerations such as relative volatilities and how the equipment is operated. However, S is can be assigned any value between 0 and 1.

Rigorous Models

The models described in the previous have two advantages. They are both simple and linear. Thus they are well suited to calculating the material balances for entire processes. However, beyond conserving mass, these models are not very realistic. Their model parameters must be known from other sources.

This presents a problem in doing accurate flowsheet material balances. In order to get the model parameters for the linear material balance calculations, we must do separate rigorous calculations for the individual items of equipment. To do this we must know the feed streams to each item of equipment for which we must do the flowsheet material balances. Thus we have a situation in which we must assume values of all of the parameters, do the flowsheet material balances using the linear models, then evaluate the parameters from the rigorous models for the individual items of equipment. It the assumed and calculated values of the parameters do no agree within some reasonable tolerance, we must repeat the procedure.

The better way is to use rigorous equipment models directly in the flowsheet material balance calculations. Since most rigorous models are nonlinear and difficult to solve in their own, this approach is not amenable to hand calculations. It is possible with a spreadsheet but considerable effort, skill, and time are required.

The way that has been adopted for doing rigorous flowsheet calculations is to use a steady-state flowsheet simulation program.

Steady-state Material Balance Calculations

There are a number of techniques that have been developed for the solution of steady-state flowsheet performance equations. In general, this is an exercise in the numerical solution of a set of algebraic (generally nonlinear) equations for which there are many algorithms and computer codes available. One approach is to write out all the equations, specify enough parameters so that the number of unknown variables equals the number of equations, and use the equation solver of choice. Indeed, this is the approach taken in many of the texts on material and energy balances. There is nothing wrong with it. However, it tends to obscure the underlying physical significance of the problem, particularly where recycles are involved. Instead we will look at some of the techniques that have been developed specifically for solving the flowsheeting problem.

Sequential Modular

One technique for solving the material balances for an entire process (although not the only one) is called Sequential Modular. In this technique the material balances for a entire process are solved one module (process block) at a time.

Let us first consider the process shown schematically in figure. There are three process operations or process blocks. The exact nature of each is not important at the moment. However, it is assumed that block equations (Eqns. $X_{out1} = G_{J,1}[X_{in1}, X_{in2}, ..., X_{inn}; P_1, P_2, ..., P_m]$, $X_{out2} = G_{J,2}[X_{in1}, X_{in2}, ..., X_{inn}; P_1, P_2, ..., P_m]$,..., $X_{outn} = G_{J,n}[X_{in1}, X_{in2}, ..., X_{inn}; P_1, P_2, ..., P_m]$) can be solved for each block in the process. In other words, if all of the input stream vectors and the parameter vector are known, then the output stream vectors for the block can be computed via a well defined procedure. (For the linear models of Eqns. $f_{i,out2} = (1 - S)\, f_{i,in}$ through $f_{i,out1} = S\, f_{i,in}$, the computations are simple and direct. For the nonlinear models used in more realistic block characterizations, the computational procedure may require a trial-and-error or iterative algorithm.)

So, the ground rule for direct sequential modular material balance calculations, the output streams can be calculated if the input streams and the block parameters are known.

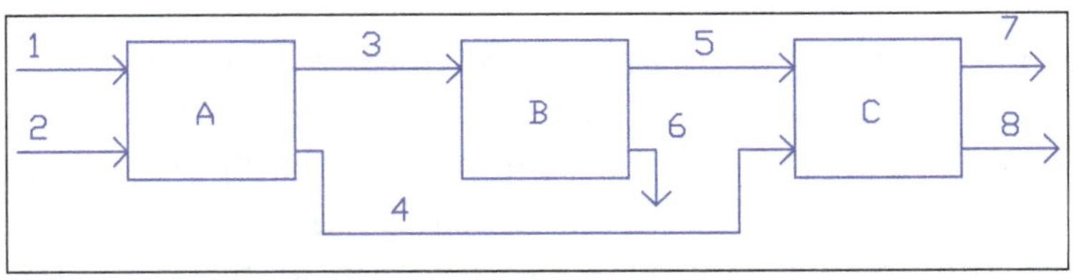

Sequential Process Flow.

Thus, in the three-block process shown, one can start by calculating Block A to determine Streams 3 and 4. Then Block B can be calculated to give Streams 5 and 6. Finally Block C can be calculated to give Streams 7 and 8.

Now let us consider a second process, one with recycle, as shown in figure. If the calculations are to be started at Block A, there is a problem. Stream 2 is not known since it is an output stream from Block B that has not been calculated yet. Suppose we decide to start with Block B instead. Again there is a problem; Streams 3 and 4 are unknown. A similar problem occurs if we attempt to start with Block C.

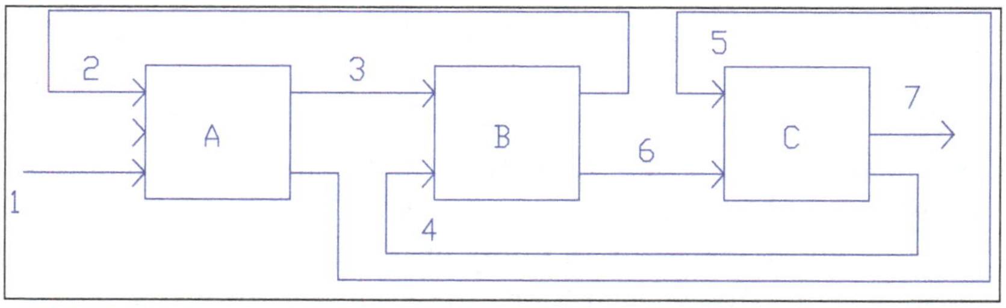

Recycle Process Flow.

So how can we possibly calculate this process? Only by agreeing to guess values for the unknown input streams to each block in the process. Suppose now we start with Block A. Stream 2 must be guessed. To calculate Block B another stream, Stream 4, must be guessed. Block 5 can be calculated without having to guess any further streams.

Once all the blocks have been calculated, we will have computed values for all the streams we originally guessed. If the computed values agree with the guessed values within some acceptable tolerance, then the material balance for the process has been solved. If not, the whole procedure has to be repeated with new guesses. One strategy is to use the previously computed values for the next round of guesses.

This technique for solving the overall flow sheet material balance problem is known as the sequential modular approach. We calculate each unit operations block or module in the process sequence, providing initial guesses of unknown recycle streams where necessary. New values of the guessed (or tear) streams are produced as a result of each pass through the process sequence calculations. When the differences between successive guessed values become sufficiently small, the procedure is considered to be converged.

The large majority of commercially available steady-state process simulators use the sequential modular approach.

Simultaneous

The sequential modular approach can generally be made to converge, even for difficult problems. However, it tends to be inefficient. Made of the unit operations models have internal iterative procedures just to solve for their output streams as functions of their input streams and operating parameters. Embed these iterative calculations within the sequential modular flowsheet calculations and one has a massive loops-within-loops calculation with the potential to be very inefficient.

this procedure is straightforward if all of the equations are linear. If they are not, an appropriate algorithm such as Newton-Raphson must be employed to solve large sets of nonlinear equations. This approach has been taken in several simulators developed in academia. SPEEDUP, developed at Imperial College, and ASCEND, developed at CarnegieMellon, are two of the more advanced of this type of simulator. SPEEDUP is currently available commercially through Aspen Technology, Inc.

It is not the intent of these notes to go into the pros and cons of the simultaneous approach versus the sequential modular. Suffice it to say that neither approach is without its drawbacks and difficulties. We note, however, that the sequential modular approach has been much widely used in commercial simulators than the simultaneous.

Design Specifications

The calculations discussed so far are, in process control terminology, for processes operated in open loop. All the input streams and block model parameters are specified and all of the block output streams are then calculated. We have no way of knowing beforehand what the values of the output stream vectors will be.

Usually, the specification of the performance that the process must achieve will involve selected variables in the output stream vectors. The production rate, i.e., the flow rate of the product stream must meet the capacity specification. This stream must also meet the product purity specifications. The composition of streams being discharged to the environmental must meet emission specifications. Many other specifications such as reactor temperature and pH, flash drum vapor flow rates, and fractional recoveries in separators must be met. This can only be done by adjusting feed stream flow rates and block model operating parameters. In some process simulators, these are known as design specifications. In fact, they have the form of feedback control loops.

There are many techniques for incorporating design specifications into the flowsheet calculations. All commercially available flowsheet simulators provide the means to do so. In sequential modular simulators, the design specifications are handled as control loops around the open loop simulation, which adjust selected parameters to satisfy the design specifications. One of the advantages of simulators using the simultaneous approach is that the design specifications can be added directly to the other equations being solved.

Optimization

Quite often one is not just interested in a solution to a flowsheet material balance problem but one that is best in some sense, usually economic. Finding such a solution by adjusting selected input and block model parameters is known as optimization. Most commercially available simulators have optimization capability. Many spreadsheets also have some optimization capability such as a linear programming solver.

Optimization is beyond the scope of these notes. We will, however, look at a specific type of optimization that arises in blending problems. The optimization technique we will use is linear programming.

Ad Hoc Methods

We term the computational method employed as ad hoc when one attempts to solve a flowsheet material balance by starting with design specifications and working backwards through the various block models and even the entire process. For flowsheets of any complexity, this a difficult approach since flowsheet calculations tend to be quite stable and well-behaved if one calculates block output streams from block input streams but notoriously unstable if one attempts to do the reverse.

Another characteristic of ad hoc methods is the use of overall material balances around two or more items of equipment, either on a component or total flow rate basis. The use of overall material balances is to be avoided. Use component material balances around individual items of equipment.

For simple flowsheets, particularly when one is just interested in the input-output structure, ad hoc methods can be used to estimate raw material requirements based on reactor selectivities and separator efficiencies. The problem is that each material balance solution is a special case. Change the form of a design specification and the entire procedure must be revised.

One is advised to use the linear material balance approach, at least initially, since it is straightforward in its application to almost any problem. If the calculations are done using a spreadsheet, then adjusting inputs and parameters to meet design specifications can be done by trial-and-error if there are not too many.

2

Sub-disciplines of Chemical Engineering

There are three major sub-disciplines within chemical engineering. These are biochemical engineering, electrochemical engineering and process engineering. This chapter closely examines the key concepts of these sub-disciplines of chemical engineering to provide an extensive understanding of the subject.

Biochemical Engineering

Biochemical Engineering is a branch of Chemical Engineering or Biological Engineering. It is a branch of engineering that deals with the design and construction of unit processes that involve biological organisms or molecules.

Biochemical engineering is usually defined as the extension of chemical engineering principles to systems using a biological catalyst to bring about desired chemical transformations. It is often subdivided into reactor design and downstream separation.

Despite the appearance that it is a new discipline, biochemical engineering as such dates back to the turn of the century, when certain rudimentary principles were used in the biological treatment of wastewater and in the production of acetone and butanol for explosives. Furthermore, fermentation ethanol was one of the basic raw materials for the chemical industry until the Second World War. The modern genesis of the discipline dates back to the early 1940s when penicillin was required in vast quantities.

The recent development of biochemical engineering is the direct result of advances in molecular biology (e.g., recombinant DNA, tissue culture, protoplast fusion, monoclonal antibodies) and protein engineering. Biochemical engineers work in a wide cross-section of industry including; pharmaceuticals, food, fine chemicals, wastewater treatment, mining and energy. Hence, biochemical products vary from high volume, low coast to very low volume and extremely high cost (e.g., Factor 8, the blood clotting protein costs around $\$10^8$/kg!). Figure details this wide variation in cost.

Biochemical engineering involves the use of a catalyst, either an enzyme or whole cell, which is either freely suspended in an aqueous medium or "immobilized" in a gel or attached to a solid surface. In contrast to an enzyme, whose activity decreases over time, cells are autocatalytic ($dx/dt = \mu x$, where x = cell concentration and μ = growth rate, t^{-1}) and can grow exponentially with excess nutrients. Reactor conditions are typically: a pH of 6–8; a temperature of 15–16 °C; atmospheric pressure; and a cell concentration of 2–40 g/l. Reactors can be operated in either batch mode (most common in the pharmaceutical industry), typically over 24–48 hours, or continuously (e.g., wastewater treatment). Reactors commonly used vary from CSTRs (continually-stirred tank reactors), through plug flow (whose efficiency is similar to CSTRs due to back mixing), to specialized reactors which immobilize the cells or enzymes to prevent them from being washed out of the reactor due to their small size and density (SG = 1.04). These types of reactors essentially "de-link" the hydraulic retention time from the cell retention time, thereby enabling process intensification to occur.

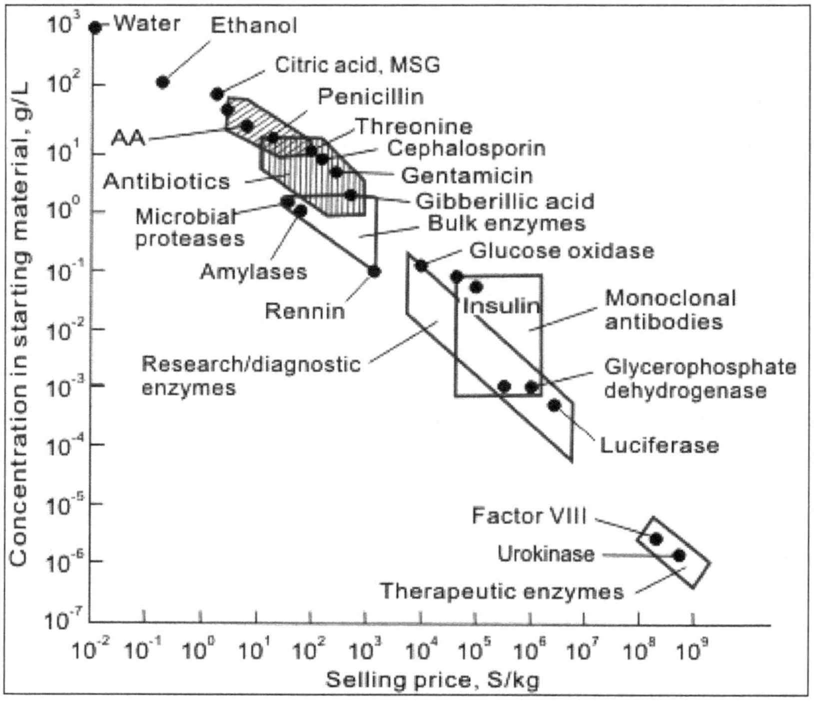

Concentration in the broth versus final selling price.

The downstream separation of biological products is one of the major challenges in biochemical engineering. These products are often present in very low concentrations (100 mg/l), and are extremely labile to mechanical shear and extremes in pH and temperature. The difficulty of the separation process depends very strongly on the product itself and whether it is the biomass, an extracellular product excreted by the cell, or an intracellular protein or inclusion body. Since most fermentation broths are extremely complex mixtures of a variety of organic and inorganic constituents and most products have to meet extremely stringent standards in terms of purity — especially the therapeutics — recovery and concentration can often involve 10–15 unit operations. This can lead to quite low overall recoveries (10–20%); hence, separation can constitute as mach as 70% of the overall product cost. The separation techniques used include those based on size, diffusivity, charge, surface activity, density and polarity, and often include techniques rarely seen in the chemical process industry.

The modeling of most biochemical engineering processes is still fraught with difficulties, although simple *enzymatic reactions* can be represented by the theoretically derived Michaelis-Menten relationship:

$$V = \frac{V_{max}\, c}{K_s + c}$$

where,

v = rate of reaction, t^{-1};

v_{max} = maximum rate of reaction, t^{-1};

K_m = Michaelis-Menten, or half rate constant, ML^{-3};

c = concentration of substrate, ML^{-3}.

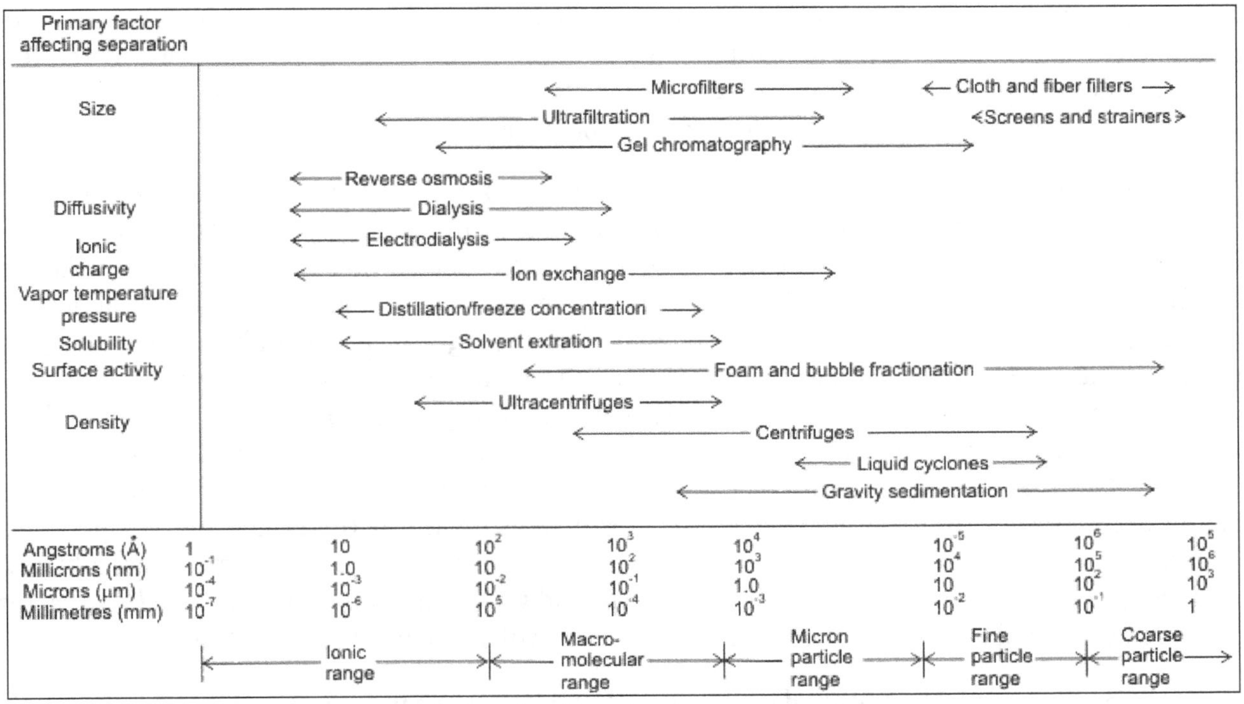

Ranges of applications of various unit operation.

However, the modeling of whole cell reactions is far more complex since cell metabolism consists of many sequential reactions (10^3–10^4), any one of which may be rate limiting at any point in time. In addition, the release of product may be kinetically associated with cell growth ("growth associated"), or occur after growth has virtually ceased ("secondary metabolite"), or be a complex mixture of both. In modeling growth kinetics, it is possible to take two approaches: using an "unstructured" model, which takes an uncritical "black box" stance; or developing a "structured" model, which is based more on the basics of cell metabolism. Due to its simplicity and relative success, the most commonly-used cell growth model is the unstructured one developed empirically by Monod:

$$\mu = \frac{\mu_{max}\, c}{K_s + c}$$

where,

μ = rate of cell growth, t^{-1};

μ_{max} = maximum rate of cell growth, t^{-1};

K_s = half rate constant, ML^{-3};

c = concentration of rate limiting substrate, ML^{-3}.

While empirically derived, the equation has obvious similarities to the Michaelis-Menten equation for enzyme kinetics.

Bioreactor

A bioreactor provides a controllable environment enabling the biological, biochemical and biomechanical requirements to manufacture engineered product. As the bioreactor aims to create a desired biological product, it is important to closely monitor the reaction parameters like internal and external mass transfer, heat transfer, Fluid velocity, shear stress etc.

Bioreactors can broadly be defined as a vessel, deployed to utilize the activity of a biological catalyst to achieve a desired chemical transformation. Bioreactor generally provides a biomechanical and a biochemical environment that controls nutrient and oxygen transfer to the cells and metabolic products from the cells. It could also be defined as an engineered device designed for optimal growth and metabolic activity of the organism through the action of biocatalyst, enzyme or microorganisms and cells of animal or plants. The raw material could be an organic or an inorganic chemical compound or even complex material. The product of conversion may include Baker's yeast, single cell protein, starter cultures, animal feed etc. or primary metabolites (e.g. amino acids, organic acids, vitamins, polysaccharides, ethanol, etc.) and secondary metabolites (e.g. antibiotics etc.). Bioreactors can be used for bioconversion or biotransformation products (steroid biotransformation, L-sorbitol etc.), enzymes (amylase, lipase, cellulase etc.), recombinant products (some vaccines, hormones such as insulin and growth hormones etc.). Varied bioreactor designs have been developed to cater to a wide array of substrate products and biocatalysts.

Bioreactors differ from conventional chemical reactors to the extent that they support and control biological entities. As the organisms are more sensitive and less stable than chemicals, bioreactor systems must be robust enough to provide a higher degree of control over process upsets and contaminations. The bioreactor conditions should be favourable for the living microorganisms to exhibit their activity under defined conditions. This calls for a series of special features in the reaction engineering of biocatalytic processes. Maintaining the desired biological activity and minimizing undesired activities are certain challenges as biological organisms, by their nature, would mutate and hence alter biochemistry of the reaction or physical properties of the organism.

The term bioreactor is often used synonymously with fermenter, which is a type of bioreactor using a living cell as the biocatalyst. Fermentation is referred to the growth of microorganisms on food, under either aerobic or anaerobic conditions. Fermenters are made up of glass, glass exotic alloys, stainless steel, glass-lined steel, plastic tanks equipped with gauges. These are used for the growth of specialized pure cultures of bacteria, fungi and yeast, production of enzymes and a wide spectrum of fermented products.

The sizes of the bioreactor can vary widely from the microbial cell (few mm^3) to shake flask (100-1000 ml) to laboratory scale fermenter (1 – 50 L) to pilot level (0.3 – 10 m^3) to plant scale (2 – 500 m^3) for large volume industrial applications. There are several aspects of biotechnological processes, which require special attention in designing a bioreactor. The reaction rate, cell growth, and process stability depend on the environmental conditions in the bioreactor. The bioreactor's conditions like gas (i.e. air, oxygen, nitrogen, carbon dioxide) flow rates, temperature, pH and dissolved oxygen levels and agitation speed/circulation rate, foam production, etc. need to be closely monitored and controlled.

Bioreactor Design and Operations

A good bioreactor design should address improved productivity, validation of desired parameters towards obtaining consistent and higher quality products in a cost effective manner. The design and mode of operation of a bioreactor depends on the production of organism, optimum conditions required for desired product formation, product value and its scale of production. The effective bioreactor is to control and positively influence the biological reaction and must prevent foreign contamination. The capital investment and operating cost are also important factors to be considered in bioreactor design. During the fermentation, monoseptic conditions, optimal mixing with low, uniform shear rates should be maintained throughout the process. Aculture can be aerated by one, or a combination, of the following methods: surface aeration, direct sparging, indirect and/or membrane aeration, medium perfusion, increasing the partial pressure of oxygen and increasing the atmospheric pressure.

Adequate mass transfer (oxygen), heat transfer, clearly defined flow condition and appropriate feeding of substrate avoiding under or overdosing would need to be maintained in a bioreactor. Proper supply of suspension of solids, sufficient substrate, salts for nutrition, vitamins etc. should be ensured with water availability and oxygen (for aerobic processes). Gas evolution product and by-product removal need to be taken care of. The attributes of a bioreactor should comply with design requirements such as sterilization, simple construction and measuring, process control devices, regulating techniques, scale-up, flexibility in operations, compatibility with upstream and downstream processes, antifoaming measures etc. are essential factors.

The basic features of a bioreactor include headspace volume, agitator system, oxygen delivery system, foam control, temperature & pH control system, sampling ports, cleaning and sterilization system and lines for charging & emptying the reactor. These are briefly described as follows.

Headspace volume: The working volume of a bioreactor is the fraction of its total volume taken up by the medium, microbes, and gas bubbles and remaining volume is called the headspace. Generally, the working volume will be ~70-80% of the total reactor volume. This, however, depends on the rate of foam formation during the reactor.

Agitator: This system consists of an external power drive, impeller and the baffles for intense mixing and increased mass transfer rates through the bulk liquid and bubble boundary layers. It provides enough shear conditions required for breaking up of bubbles (srmuni.ac.in). Most microbial fermentations use a Rushton turbine type impeller.

Air delivery: This system consists of a compressor, inlet air, sterilization system, air sparger and exit air sterilization system to avoid contamination.

Foam control: This system is an essential element of bioreactor as excessive foam formation leads to blocked air exit filters and builds up pressure in the reactor.

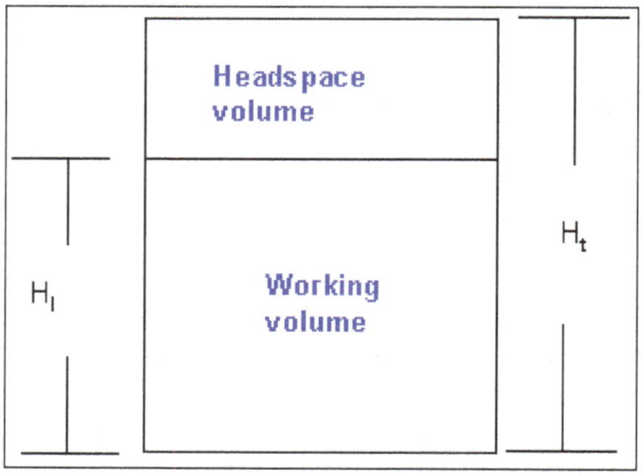

Headspace volume = Ht (Total volume of bioreactor) - Hl (Working volume).

Temperature control: This system involves temperature probes, heat transfer system (jacket, coil). Heating is provided by electric heaters and steam generated in boilers and cooling is provided by cooling water produced by cooling towers or refrigerants such as ammonia.

pH: It control system uses neutralizing agents to control pH; these should be non-corrosive, non-toxic to cells when diluted in the medium. Sodium carbonate is commonly used in small scale bioreactor.

Sampling ports: They are used to inject nutrients, water, salts etc. in bioreactors and also for collecting samples.

Cleaning and sterilization: This system is important to avoid contamination. Thermal sterilisation by steam is preferred option for economical and large-scale sterilizations of equipment. Sterilization by chemical substances is generally preferred for heat-sensitive equipment. Sterilization is carried out by radiation by uv for surfaces and x-rays for liquids and also by membrane filters having uniform microspores and depth filters with glass wool. Charging & emptying lines are used for input of reactants and for withdrawal of products in the bioreactor.

Bioreactors – An Insight into Mass Transfer

Mass transfer, referred as movement of molecules between phases controls or in influences the conversion rate of the reactions. In a bioreactor, components for mass transfer may typically include oxygen, electron acceptor, total organic carbon (TOC), chemical oxygen demand (COD), biomass, ammonium and nitrate and macro nutrients. The transport of dissolved oxygen (DO) in a bioreactor occurs in three regions as (i) bulk fluid phase of the bioreactor (global mass transfer), (ii) from the bulk to the surface of the aggregated cells (internal mass transfer), (iii) through the aggregated cells (external mass transfer).

The mass transfer coefficient is evaluated by using an oxygen sorption method. The interfacial mass transfer area is determined on the basis of its measured bubble size distribution. The liquid-side mass transfer coefficients can be calculated from the volumetric mass transfer coefficients and the interfacial mass transfer area found. The liquid-side mass transfer varies with the superficial gas velocity as dictated by varying bubble rising parameters. Static liquid height improves gas-liquid mass transfer rate for increased interfacial mass transfer area. Bubble size due to type of gas distributor used also governs the mass transfer rate.

The two-film theory as proposed by Lewis and Whitman in 1924 attempts to quantity mass transfer from gas to liquid phase of the target compounds in biological systems. According to Henry's law, this model involves two phase s with different concentrations, which are not in equilibrium. With the equilibrium existing only at the gas–liquid interface, the target compound moves towards or away from the interphase. Velocity of such movement depends on the properties and type of compound s of the two phases i.e. gas & liquid. Such velocities are expressed by the mass transfer rate coefficients. Overall mass transfer coefficient is a combination of mass transfer coefficients of gas phase, liquid phase and biofilm as given in the following equation:

$$\frac{1}{k_{overall}} = \frac{1}{k_G} + \frac{1}{k_L} + \frac{1}{k_B}$$

Where,

$K_{overalll}$ - Overall mass transfer coefficient

k_G - Mass transfer rate coefficient of gas phase

k_L - Mass transfer rate coefficient of liquid phase

k_B - Mass transfer rate coefficient of the biofilm

The mass transfer coefficients depend on medium and physiochemical properties, internal reactor system characteristics and the operating conditions. In suspended type reactors (e.g. airlift, bubble columns, and stirred tanks), the above equation can be applied considering k_B as the resistance due to the water film around the cell. Considering resistance to mass transfer in the gas phase and the biofilm to be negligible, the overall volumetric mass transfer rate R from the gas phase to the aqueous phase may be defined in the following equation:

$$R = k_{La}\left(\frac{C_G}{H\text{-}C_L}\right) + \frac{DAL}{\delta_{film}}\, a\left(\frac{C_G}{H\text{-}_{CL}}\right)$$

Where,

D_{AL} - Gaseous pollutant diffusivity in the liquid ($m^2\ s^{-1}$)

H - Henry coefficient (dimensionless)

δ_{film} - Liquid film thickness (m)

C_G -Pollutant concentrations in gas phase

C_L -Pollutant concentrations in liquid phase

k_{La} – Volumetric coefficient (s^{-1})

a - Specific interfacial area ($m^2\ m^{-3}$) between the gas and liquid phase

Internal mass transfer rates depend on a combination of convection mechanisms, scaffold's structure and porosity, diffusion rate whereas external mass transfer rates depend on hydrodynamic conditions in a bioreactor.

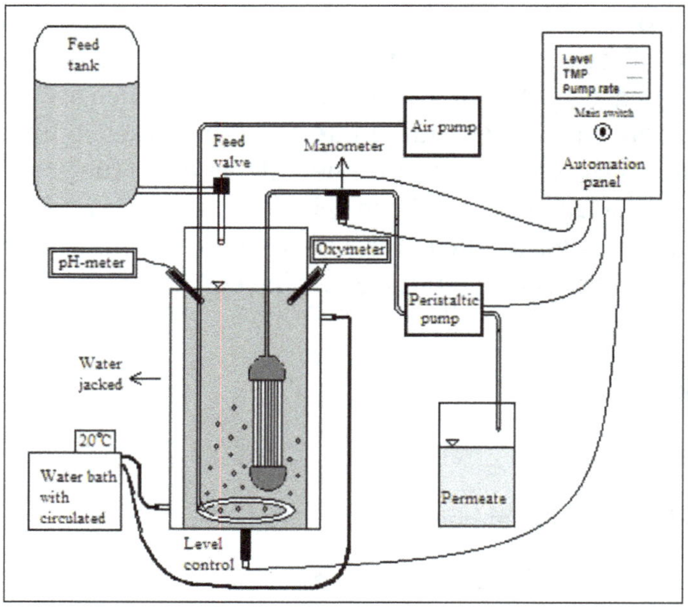

Schematic Representation of Bioreactor-1.Engine regulator 2.engine 3.inoculum 4.carbon source 5.anti-foam 6.anti-foam controller 7.thermometer 8.air exhaust system with filter and condenser 9.sampling 10.water bath 11.temperature controller 12.air filter.

Types of Bioreactor

There are mainly three types of reactions involved in fermentation process i.e. batch, continuous and semi-continuous or fed-batch depending on the feeding strategy of the culture and the medium into the bioreactor. Traditional batch stirred tank reactors (STRs) and continuously stirred tank reactors (CSTRs) have existed for centuries and are still widely adopted in the chemical and bioprocessing industry for production due to their simplicity. Other bioreactors, which have special design and operational attributes are photo-bioreactors, rotary drum reactors, mist bioreactor, membrane bioreactor, packed & fluidized bed bioreactors, bubble column & air lift bioreactors etc. These have been developed to cater to application specific processes.

Batch Process

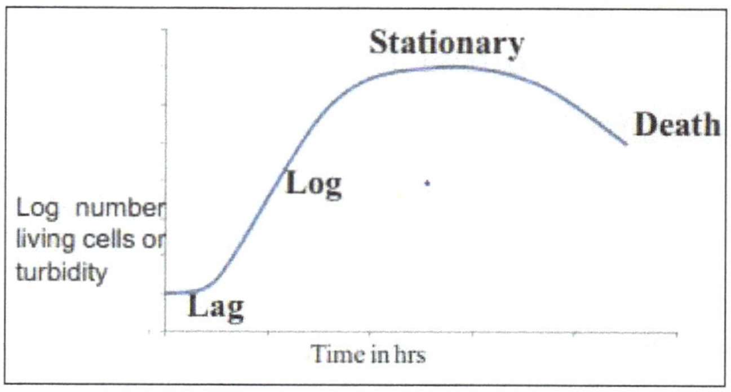

Growth Curve.

In the batch process, after sterilisation, the sterile culture medium is inoculated with microorganisms. During this reaction period, cells, substrates including the nutrient salts, vitamins and concentrations of the products vary with time. The fermentation is allowed to run for a predetermined time and the product is harvested at the end. To promote aerobic cultivation, the medium is aerated to provide a continuous flow of oxygen. Gaseous by-products such as CO_2 are removed; aeration and gas removal processes take place semi continuously.

- Lag phase: The growth of microbial population when it is inoculated with a fresh medium starts after a certain period of time called lag phase.
- Log or Exponential phase: In this phase, the microbial cell numbers double per unit time period. When the cell number from such a reaction is plotted on logarithmic scale as function of elapsed time, a curve is obtained with a constantly increasing slope.
- Stationary phase: In stationary phase there is no net increase or reduction in cell number. The cell functions such as energy metabolism and some biosynthetic processes go on.
- Death phase: The cells may start dying if the incubation is continued after the bacterial population attains the stationary phase. Cells may die due to cell lysis, which is a much slower process than the growth phase.

Batch Bioreactor Design

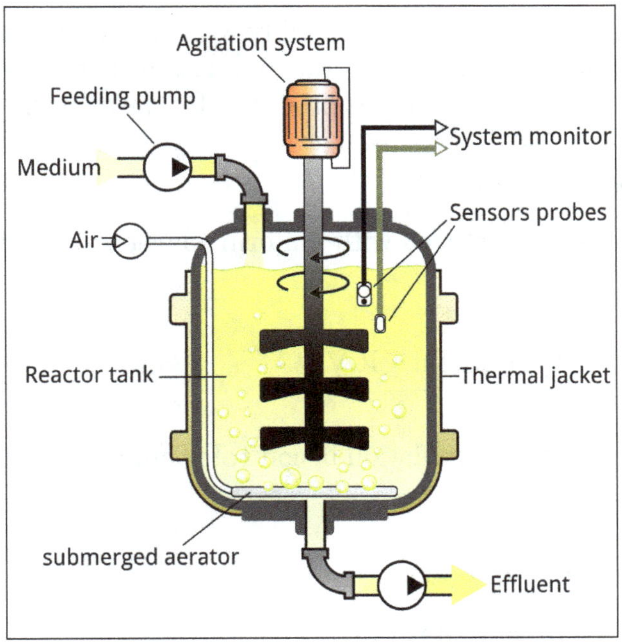

Bioreactor.

Batch bioreactors comprise of single tank capable of carrying out sequence of reactions and are easy to operate. The tank is equipped with an agitator (stirred tank reactor – STR) to mix the reactants along with integral heating and cooling system. Buffer solution or pH controller is used to control pH of the reactant. These vessels may vary in capacity from less than 1 litre to more than 15,000 litres. Liquids and solids are usually charged via inlets in the top cover of the reactor. Vapours and gases also discharge through connections in the top. Usually liquids are removed from the bottom. STRs generally jacketed for steam heating or cooling requirements and are equipped with baffles and round sparger for aeration.

The impeller in STRs is connected to an external motor, which drives the stirrer system. The agitator assembly, including the seal, is a route of contamination and hence the shaft has to pass into the bioreactor through a set of aseptic seals. The impellers contribute to mixing and dissolution of the required atmospheric oxygen into the aqueous phase, and maximize the interfacial area between the gaseous and aqueous phase. The design of the impeller blades, speed of agitation and the depth of liquid determines the effectiveness of agitation. The important variables, which affect mixing and mass transfer rates are number and types of stirrer, speed of stirrer and the flow rate of gas used.

These reactors are preferred for low-volume & high-value products, particularly if many sequential operations are employed to obtain product yields. These reactors are also used when multiple products are produced in the same equipment or when continuous flow is difficult, as in case of highly viscous or sticky solids-laden liquids. STRs are used for homogenization, suspension of solids, dispersion of gas-liquid mixtures, aeration of liquid and heat exchange. They are most common types of aerobic bioreactors in use today; they may feature a specific internal configuration designed to provide specific circulation pattern. They can be used with a variety of microbial species and widely adopted for microorganisms, fermentation and plant cell culture. Nutrient concentration pH and amount of dissolved oxygen can be controlled within this type of bioreactors.

Advantages of batch reactors include more flexibility with varying product systems and with reduced risk of contamination or cell mutation, due to a relatively brief growth period with lower capital investment as compared to continuous processes for the same bioreactor volume.

Mass Balance for Batch Reactors

The batch reactor is assumed well stirred with uniform concentration distribution across the reactor. Hence,

$$\frac{\ddot{u}(VC)}{\ddot{u}} = Q\ in\ .C\ in - Q\ out.C\ out + R.V$$

Where d(VC)/dt = Rate of mass accumulation in control volume

Q in = flow rate into the system

Q out = flow rate out of the system

C = Concentration of stream/substrate

R = Rate of reaction

V= Volume of the stream/substrate

The inflow and outflow stream rates are zero:

Q in – Q out = 0

Hence,

(VC)/dt = RV(if reactant volume changes significantly) or

d(C)/dt = R (if reactant volume remains constant)

R= k. C where k = rate constant, C= concentration

Continuous Process

For a bioreactor on continuous mode operations, fresh medium is continuously added and the products, along with the culture are removed at the same rate, thus maintaining constant concentrations of nutrients and cells throughout the process. Continuous process is frequently used for high-volume production; for reactions using gas, liquid or soluble solid substrates; and for processes involving microorganisms with high mutation-stability. Typical end products include vinegar, baker's yeast and treated wastewater. Chemostat is a common example of continuous process reactor.

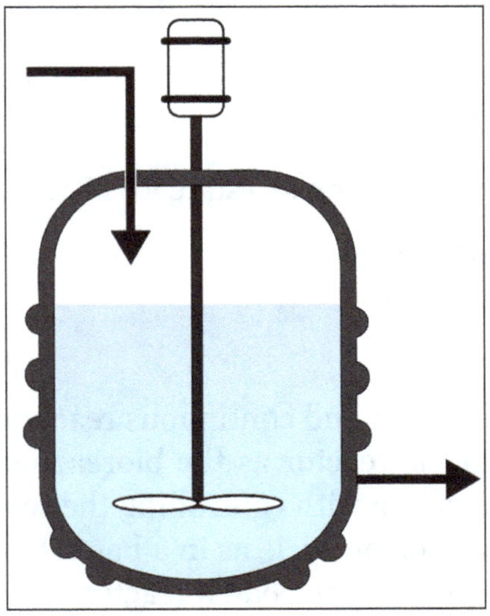

Continuous reactor.

Continuous Bioreactor Design

The reactants are well mixed in a continuous stirred-tank reactor. They are also known as vat- or backmix reactor. The characteristic feature of continuous bioreactor is a perpetual feeding process. Liquid or slurry stream is continuously introduced and liquid contents are continuously removed from the reactor. In practice, mechanical or hydraulic agitation is required to achieve uniform composition and temperature. A culture medium that is either sterile or comprises of microorganisms is continuously fed into the bioreactor to maintain the steady state. The reaction variables and control parameters remain consistent, establishing a time constant state within the reactor. The result is continuous productivity and output.

CSTR requires large volume reactors to obtain desired conversions. Configuration of such reactors is widely used in industrial applications and in wastewater treatment units (i.e. activated sludge reactors). By automating the process, there could be reduced labour expense and time saving in Filling, emptying and sterilizing the reactor with reduced toxicity risks. CSTRs can yield consistent product quality to invariable operating parameters. However, CSTRs are more energy consuming due to the presence of mechanical pumps.

Mass Balance for CSTR

The material balance for this reactor gives:

$$\frac{d(VC)}{dt} = \text{Qin. Cin - Qout. Cout + R.V}$$

If the reactor volume is constant and flow rates of the in flow and out flow streams are the same, then;

$$\frac{d(C)}{dt} = \frac{1}{\tau}\left(\text{ Cin - Cout + R}\right)$$

This parameter $\tau = \dfrac{v}{q}$ in it is called the mean residence time of the CSTR.

The steady state of the CSTR is described by setting the time derivative in the expression, d(VC)/dt = o

Q in . C in - Qout . Cout + R.V= o

Conversion of reactant 'X' is defined for a steady-state CSTR as follows:

X = (Qin. C in - Qout. Cout) / Q in .C in

Fed-batch Process

The process uses a combination of batch and continuous reactions. In this process additional nutrients are added progressively to the reactor as the bioreactions are underway so as to obtain better yields and higher selectivity along with controlling the reaction temperature. The products are harvested at the end of the production cycle as in a batch bioreactor. Semi-batch reactors are stabler and perform safer operations than in a batch reactor.

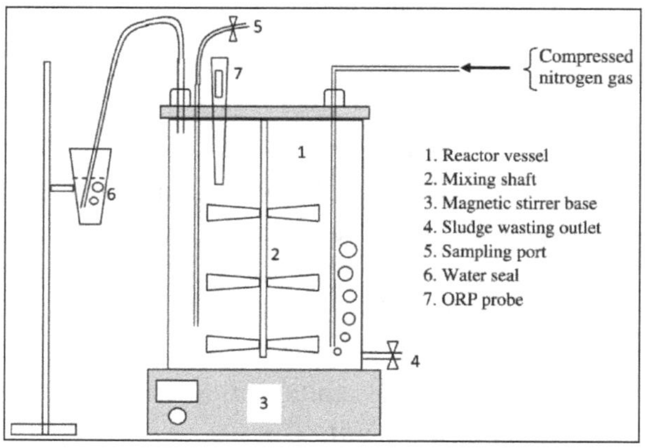

Fed-batch bioreactor.

Special Purpose Bioreactors

Plug Flow Reactor

Plug flow reactors are also referred to as a tubular or piston- flow reactor. It is a vessel, through which the flow is continuous and unidirectional in a steady state. In ideal tubular reactor, the fluids flow as if they were solid plugs or pistons, and reaction time is the same for all flowing material at any given tube cross section. The fluid is hypothesized to flow as plugs or pistons in a tubular reactor with identical reaction time over the reactor cross-section. The concentration of substrates and microorganisms vary throughout the reactor. Tubular reactors are functionally similar to batch reactor as they provide high driving force initially; this reduces as the reaction continues along the tubes.

Fluid flow in small diameter tubes could be laminar for highly viscous liquids and turbulent for gases. Turbulent flow regime for its positive influence on mixing and heat transfer is a preferred choice. The heat transfer rate can be optimised using tubes with larger or smaller diameter arranged in parallel. However, the control of temperature and heat can result in undesirable temperature gradients and which is expensive to maintain.

Table: Comparison on basis of Mode of Operation.

Mode of Operation	Advantages	Disadvantages
Batch	Simple equipment; suitable for small production volumes along with multi-product flexibility.	Downtime for loading and cleaning; reaction conditions change with time.
Continuous	High productivity; better product quality due to constant conditions; good for kinetic Studies.	Requires flow control, longevity of catalyst necessary, stability of organisms.
Semi-batch or Fed-batch operation	Control of environmental conditions e.g. substrate concentration (inhibition), induction of product formation; most flexible for selecting optimal conditions; most frequently used in biotechnological processes and in fine chemical industry.	Requires feeding strategy e.g. to keep constant temperature or substrate concentration.

Mass Balance for Plug Flow Reactor

Plug flow in a tube is an ideal- flow assumption in which the fluid is well mixed in the radial directions. The fluid velocity is assumed to be a function of only the axial position in the tube.

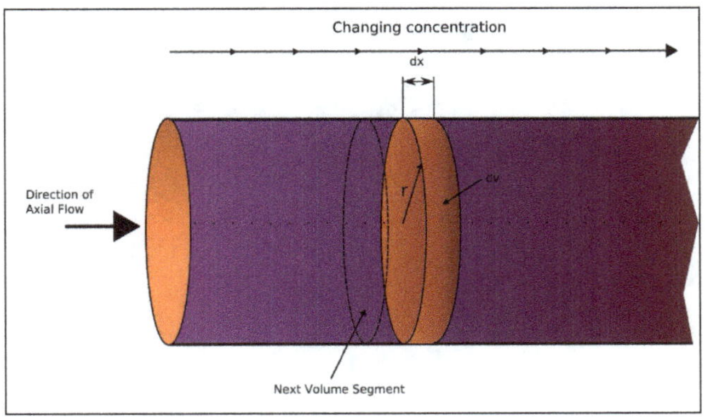

Plug flow reactor – graphical representation.

Considering the reactor cross sectional areas as 'A c' and a thin disk with infinitesimal thickness 'Δz' for the reactor volume element, the material balance for the volume element is as follows:

$$\frac{d(VC)}{dt} = Q\ in\ z.\ C\ in - Qout\ z{+}\Delta z.\ C\ out + R.\ \Delta V$$

Where $\frac{\ddot{u}(\quad)}{dt}$ = Rate of mass accumulation in control volume

Q in (flow rate into the system) = Q out (flow rate out of the system) = Q

Cin = Concentration of stream at disk inlet (z); Cout = Conc. of stream at disk outlet ($z + \Delta z$)

R = Rate of reaction

V= Reactor volume

Dividing the above equation by ΔV and taking the limit as ΔV goes to zero yields,

$$\frac{d(C)}{dt} = -\frac{dC.Q}{dV} + R$$

If the tube has constant cross section, 'Ac', then velocity, 'v' is related to volumetric flow rate by v = Q/Ac, axial length is related to tube volume by z = V/Ac,

The equation can be rearranged as:

$$\frac{dC}{dt} = \frac{-d(Cv)}{dz} + R \text{ or } R = \frac{dC}{dt} - \frac{d(Cv)}{dz}$$

Bubble Column Reactor

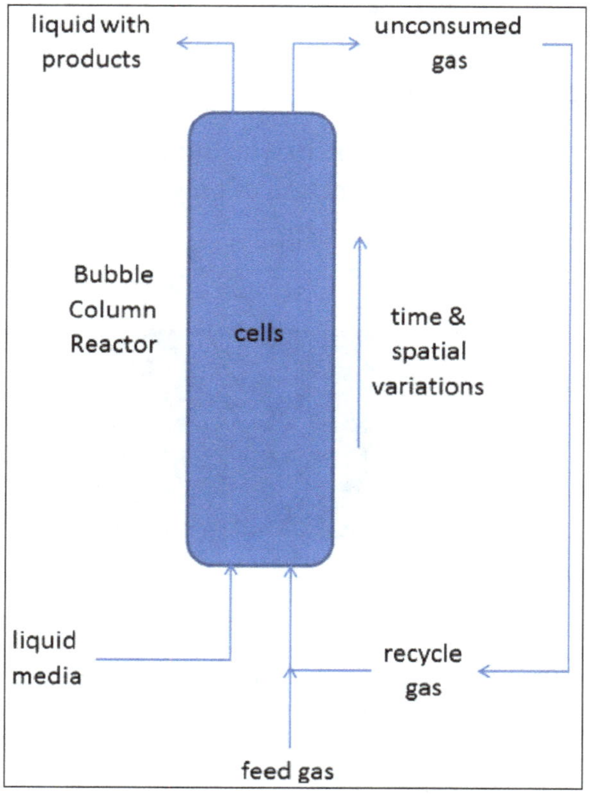

Bubble column bioreactor.

The bubble column reactor is one of the simplest types of reactors, which is easy to scale-up. The reactor comprises of a cylindrical vessel provided with a gas sparger, which pushes gas bubbles into a liquid phase or a liquid-solid suspension. For rectors handling a solid phase, the reactors are termed as slurry bubble column reactors. The reactors can be deployed to manufacture environmentally benign synthetic fuels such as methanol. The bubbles create less shear stress compared to other reactors with agitators.

Bubble column reactors have excellent heat and mass transfer characteristics. They call for little maintenance and low operating costs due to lack of moving parts and compactness. Bubble column reactors are preferred choice for high durability of the catalyst or other packing material, online catalyst addition and withdrawal ability. Bubble columns reactors are used in biochemical processes such as fermentation and biological wastewater treatment. These reactors are also used in methanol synthesis, and manufacture of other synthetic fuels which are environmentally much more advantageous over petroleum-derived fuels. These are particularly useful for hairy root culture of the plant cells. Although the construction of bubble columns is simple, efficient design and scale-up require an improved understanding of multiphase fluid dynamics and its influences. Their design depends on the three main phenomena i.e. heat, mass transfer & mixing characteristics and chemical kinetics of the reacting system Industrial bubble columns usually operate with a length-to-diameter ratio of at least 5. In biochemical applications this value usually varies between 2 and 5.

Airlift Bioreactor (ALB)

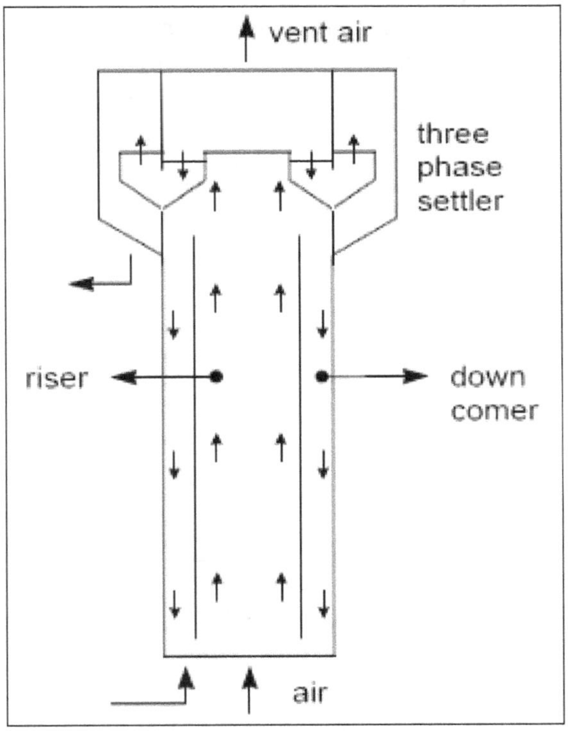

Airlift reactor.

Airlift bioreactor, also known as a tower reactor uses the expansion of compressed gas for mixing. ALBs can be used for both free and immobilized cells and are suitable for bacteria, yeast, fungi, plant, and animal cell. In these reactors, the fluid volume is divided by providing an inner draft tube for improving circulation & oxygen transfer and equalizing shear forces in the reactor. Air flows up the riser tube, forming bubbles, and exhaust gas is released from the top of the column. The degassed liquid then flows through downcomer and the product is emptied from the bottom of the tank. The downcomer tube can be designed to serve as an internal heat exchanger, or a heat exchanger can be added to an internal circulation loop. Sparging is done either inside or outside the draft tube. In absence of agitation, the reactor requires low energy making it an energy efficient system. ALBs have increased masstransfer as enhanced oxygen solubility is achieved in large tanks with controlled flow and efficient mixing with good residence time.

Packed Bed Bioreactors

The reactors necessarily constitute a bed of packings, made of polymer, ceramic, glass, natural material, and available in a variety of shapes and sizes that allows fluids to flow from one end to the other. The immobilized biocatalyst is packed in the column and fed with nutrients either from top or from bottom. Fluid comprising of dissolved nutrient and substrate flows through the solid bed. The fluid flow rate and residence time are controlled to increase or decrease substrate contact with the bed. The packed-bed compartment located either external to, or within, the reservoir of the medium.

Packed beds can either be run in the submerged mode (with or without aeration) or in the trickle flow mode. The flow velocities in the channels can be high to eliminate external mass transfer limitation in the adjacent liquid film. Simultaneously, plugging can be avoided, although at the cost of high pressure drop. Undesired properties of these reactors include poor temperature control, heat gradients, unwanted side reactions and difficulty in replacing catalyst. They also suffer from blockages and poor oxygen transfer. Changes in the bed porosity during operation alter the flow characteristics of these reactors. These are generally used in waste water engineering.

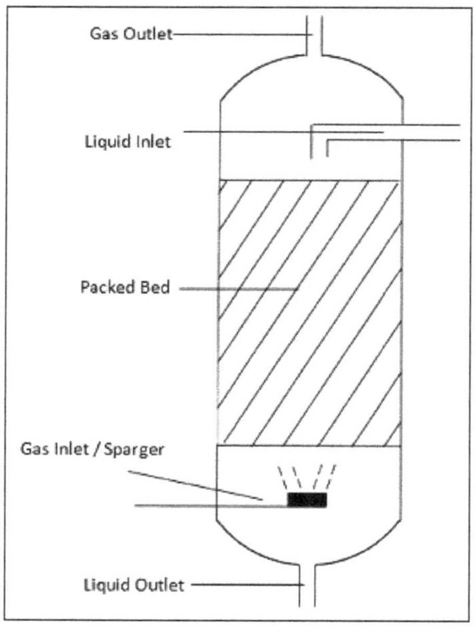

Packed bed reactor.

Fluidized Bed Reactor

Fluidized bed reactors (FBRs) constitute packed bed with smaller size particles. Thus the problems of clogging, high liquid pressure drop, channelling and bed compaction are prevented as compared to packed bed reactors. These reactors operate in a continuous state with uniform particle mixing and temperature gradients. In these reactors, the cells are immobilized small particles which move with the fluid. The smaller particle size facilitates higher rate of mass transfer, oxygen transfer and nutrients to the cells. The biocatalyst concentration can significantly be higher and washout limitations of free cell systems can be overcome. In this reactor, the cross-section area is expanded near the top to reduce superficial velocity of fluidizing liquid to a value below the terminal velocity of the particles to prevent elutriation. The efficiency of fluidized bed reactor depends on the attachment of particles that are maintained in suspension by an upward flow rate of the fluid to be treated. The particles are often called biofilm carriers and either they are inert core on which the biomass is created by cell attachment; or porous particles in which the biocatalyst is entrapped (self-immobilization).

The fluidized reactors are operated in co-current up flow with liquid as continuous phase. Usually fluidization is obtained either by external liquid re-circulation or by gas fed to the reactor .Some undesirable properties of FBRs includes increased reactor vessel size; pumping requirements and pressure drop, particle entrainment, erosion of internal components, pressure loss etc.

Bioreactors have been used for decades to produce a range of therapeutic biomolecules and other high-value products. They provide the opportunity to monitor and control environmental conditions continuously throughout the culture/reaction period along with the added benefit of maintaining a closed system. They are critical and integral part of the development of many new processes.

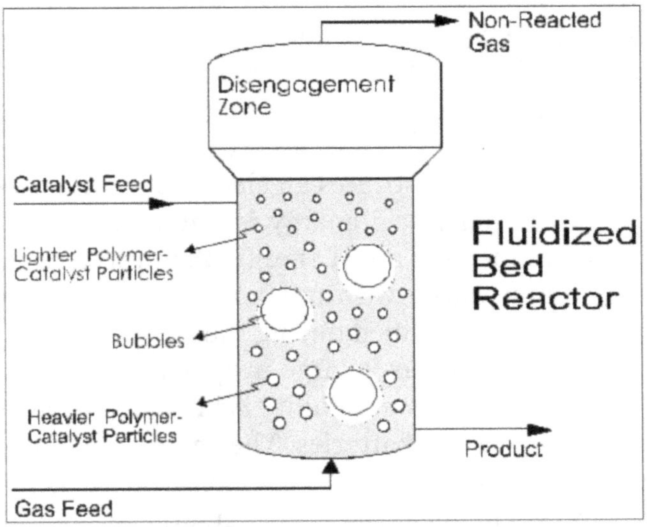

Fluidized bed reactor.

The proper selection and design of the bioreactor addressing high process efficiencies would determine the economic viability of bioprocess and its corresponding capital investment. Suitable process engineering calculation methods have been developed to give a quantitative understanding of mass transfer. Innovative methodologies for gas transfer, maintenance of pH, sensors and actuators detecting temperature, optimal feeding and cell quantification etc. are important tools for process engineering. As bioreactors are highly dependent on temperature control, it is essential to select the suitable temperature control device based on the specific requirements of each application by calculating the heat load. The type of bioreactor would depend upon the morphology of cells, shear tolerance, growth and production behaviour of the culture.

Electrochemical Engineering

Electrochemical Engineering is the branch of chemistry that deals with the chemical changes produced by electricity and the production of electricity by chemical changes.

Electrochemical engineering is the knowledge required to either design and run an industrial plant which includes an electrolytic stage for the production of chemicals or to produce an electrolytic device for the generation of power. The former involves the use of electric power for the production of chemicals and the latter the use of chemicals for the production of electric power. There are also a number of processes based on electrolysis, for example, electrochemical machining and electrophoretic painting, which clearly belong to the domain of electrochemical engineering.

Electrochemical Reaction

Electrochemical reaction refers to any process either caused or accompanied by the passage of an electric current and involving in most cases the transfer of electrons between two substances—one a solid and the other a liquid.

Under ordinary conditions, the occurrence of a chemical reaction is accompanied by the liberation or absorption of heat and not of any other form of energy; but there are many chemical reactions that—when allowed to proceed in contact with two electronic conductors, separated by conducting wires—liberate what is called electrical energy, and an electric current is generated. Conversely, the energy of an electric current can be used to bring about many chemical reactions that do not occur spontaneously. A process involving the direct conversion of chemical energy when suitably organized constitutes an electrical cell. A process whereby electrical energy is converted directly into chemical energy is one of electrolysis; i.e., an electrolytic process. By virtue of their combined chemical energy, the products of an electrolytic process have a tendency to react spontaneously with one another, reproducing the substances that were reactants and were therefore consumed during the electrolysis. If this reverse reaction is allowed to occur under proper conditions, a large proportion of the electrical energy used in the electrolysis may be regenerated. This possibility is made use of in accumulators or storage cells, sets of which are known as storage batteries. The charging of an accumulator is a process of electrolysis; a chemical change is produced by the electric current passing through it. In the discharge of the cell, the reverse chemical change occurs, the accumulator acting as a cell that produces an electric current.

General Principles

Substances that are reasonably good conductors of electricity may be divided into two groups: the metallic, or electronic, conductors and the electrolytic conductors. The metals and many nonmetallic substances such as graphite, manganese dioxide, and lead sulfide exhibit metallic conductivity; the passage of an electric current through them produces heating and magnetic effects but no chemical changes. Electrolytic conductors, or electrolytes, comprise most acids, bases, and salts, either in the molten condition or in solution in water or other solvents.

Plates or rods composed of a suitable metallic conductor dipping into the fluid electrolyte are employed to conduct the current into and out of the liquid; i.e., to act as electrodes. When a current is passed between electrodes through an electrolyte, not only are heating and magnetic effects produced but also definite chemical changes occur. At or in the neighbourhood of the negative electrode, called the cathode, the chemical change may be the deposition of a metal or the liberation of hydrogen and formation of a basic substance or some other chemical reduction process; at the positive electrode, or anode, it may be the dissolution of the anode itself, the liberation of a nonmetal, the production of oxygen and an acidic substance, or some other chemical oxidation process.

An electrolyte, prepared either by the melting of a suitable substance or by the dissolving of it in water or other liquid, owes its characteristic properties to the presence in it of electrically charged atoms or groups of atoms produced by the spontaneous splitting up or dissociation of the molecules of the substance. In solutions of the so-called strong electrolytes, most of the original substance, or in some solutions perhaps all of it, has undergone this process of electrolytic dissociation into charged particles, or ions. When an electrical potential difference (i.e., a difference in degree of electrification) is established between electrodes dipping into an electrolyte, positively charged ions move toward the cathode and ions bearing negative charges move toward the anode. The electric current is carried through the electrolyte by this migration of the ions. When an ion reaches the electrode of opposite polarity, its electrical charge is donated to the metal, or an electric charge is received from the metal. The ion is thereby converted into an ordinary neutral atom or group of atoms. It is this discharge of ions that gives rise to one of the types of chemical changes occurring at electrodes.

Electrochemical Process

Interactions of matter associated with the passage of an electric current depend upon the characteristics of the negatively charged electron. As the basic particle of electricity, the electron has an affinity for positively charged particles of matter, protons, whether in atoms, groups of atoms, or molecules. This affinity is analogous to the chemical affinity that particles exhibit among themselves. In fact, all chemical reactions result from a shift in the electron structure of atoms, and free electrons can combine with particles of matter (reduction) or be released by them (oxidation). The quantitative relationship between the free electrons of an electric current and the particles of a substance in which they cause a reaction is defined by the laws of Faraday). The substances that take part in electrochemical reactions, called electrolytes or ionic conductors.

Electrons are available in large quantities in a relatively free (mobile) state only in substances called electronic conductors, among which metals are the most important. Thus, an electron conductor must be present as a basic component of any system in which electrochemical reactions are to occur. Furthermore, the availability of electrons in a conductor is limited by energy distribution to such an extent that electrochemical reactions take place only in the immediate vicinity of the electronic conductor's surface—i.e., a few angstroms from the conductor into the solution. These reactions are, therefore, normally considered as occurring at the interface, or common boundary, between an electronic conductor, such as an electrode, and an ionic conductor of electricity, such as an electrolytic solution. Electrochemical reaction will take place, however, only to the extent that electricity can flow through such a system as a whole. To achieve this, it is necessary for the system to form a closed loop, electronically speaking.

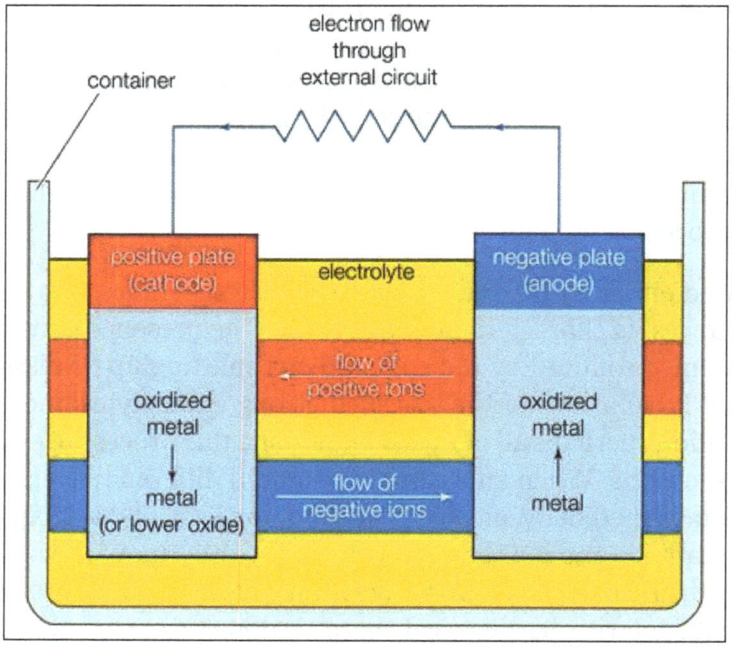

The mechanism of electron movement in an electrochemical cell.

To summarize, if at one metal-solution interface electrons are coming out of the metal, reducing a component of the solution, there must exist a second metal-solution interface where electrons are going into the metal in the process of oxidation.

The two electrodes and the ionic conductor in between (e.g., an aqueous solution of some ionized salt) represent an electrochemical cell. The process occurring in the cell as a whole is a redox process with the reduction of one species spatially separated from the oxidation of another one. As a consequence of Faraday's law, the rates of electrochemical reactions at electrodes (expressed in terms of gram moles per second per square centimetre of electrode surface) are directly proportional to the current density (expressed in amperes per square centimetre)—i.e., current flowing through the cell divided by the electrode surface area.

Sites of Electrochemical Reactions

Electrochemical reactions take place where the electron conductor meets the ionic conductor—i.e., at the electrode–electrolyte interface. Characteristic of this region, considered to be a surface phase, is the existence of a specific structure of particles and the presence of an electric field of considerable intensity (up to 10,000,000 volts per centimetre) across it; the field is caused by the separation of charges that are present between the two bulk phases in contact. For most purposes the surface phase can be considered as a parallel plate condenser, with one plate on the centre of the ions that have been brought to the electrode, at the distance of their closest approach to it, and with the second plate at the metal surface; between the two plates and acting as a dielectric (i.e., a nonconducting material) are oriented water molecules. This structure is termed the electric double layer and is illustrated in figure.

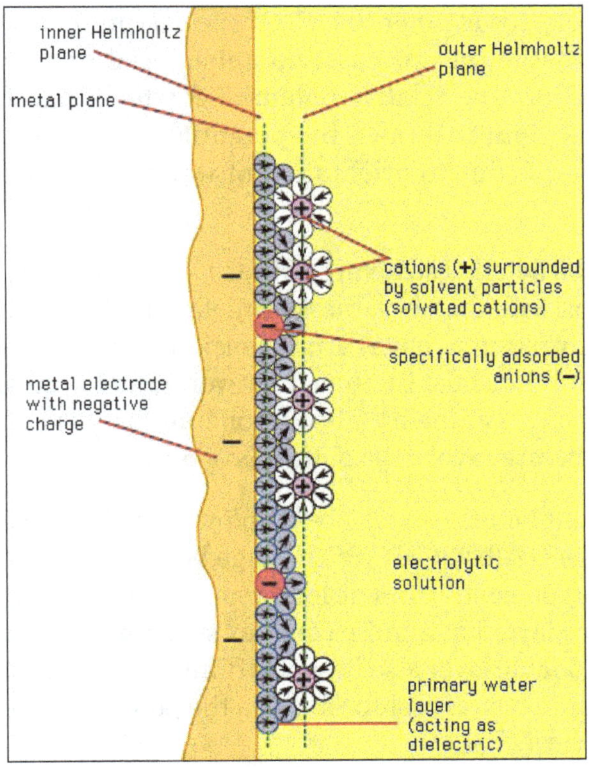

Double-layer structure and change of potential with distance from the electrode surface.

Thermal motion of the positive ions in the solution makes the condenser plate on the electrolyte side of the interface diffuse—i.e., the ions are distributed in a cloudlike way. This condition justifies the division of the potential change between the bulk of metal and the bulk of electrolyte into two parts: first, that between the metal surface and the first ionic layer at the distance of closest approach (called the outer Helmholtz plane, in which the ions are usually surrounded by solvent particles—i.e., are solvated); and second, that between the first ionic layer and the bulk of the solution, the diffuse part of the double layer. The picture is further complicated by the presence of ions in the electrode surface layer in addition to those that are present for electrostatic reasons—i.e., by the force of attraction or repulsion between electric charges. Such electrode surface layer ions are said to be specifically adsorbed on the electrode surface. Since this species of ions is attracted by the surface to a distance closer than the "distance of the closest approach" of ions, further subdivision of the inner part of the electric double layer is justified. Hence, the inner Helmholtz plane is introduced as the plane formed by the centres of specifically adsorbed ions. Adsorption of neutral molecules on the surface can also change the properties of the electric double layer. This change occurs as a consequence of replacing the water molecules, and thus changes that part of the potential (electrical) difference across the double layer that is caused by the adsorbed dipoles (water molecules that have a polarity—i.e., they behave like minute magnets—because of their hydrogen-oxygen structure, making one end of the molecule positive and the other end negative).

The absolute value of electrical potential difference, symbolized in calculation by the Greek letters delta and psi, $\Delta\Psi$, between the bulk of a metal electrode and the bulk of an electrolyte cannot be measured. Instead, the voltage of a special cell, composed of the specific electrode being studied and of an arbitrarily selected reference electrode, is normally measured; the voltage is referred to as the relative electrode potential, *E*. Of special interest is that state of the electrode at which there is no net charge (in this case, no unbalanced or extra positive, charge) at the metal side of the double layer. The relative potential at which this state is achieved is characteristic of each metal. This point is termed the potential of zero charge. At that potential, the field across the double layer is due to orientation of water molecules and other dipoles at the surface only.

Most of the knowledge of the detailed structure of the interface between a metal and an electrolyte arises from experimentation with mercury, the only metal that is liquid at ordinary temperatures; the double layer structure turns out to have surface tensions that must be measured, and this measurement is difficult with solid metals. By 1970, however, it had been shown that it is possible to measure surface tension changes at the metal-solution interface. Thus, the way to the determination of the double layer structure involving solids was opened.

Substances that are semiconductors can also be employed as electron carriers in electrochemical reactions. Semiconductors are substances which range between serving as insulators at low temperatures and as metallic-type conductors at high temperatures. In the case of semiconductors, however, the electric double layer has a more complex structure inasmuch as the condenser plate at the electrode side of the double layer also becomes diffuse. Thus, the overall potential difference between bulks of the phases in contact comprises also the potential difference between the bulk of the semiconductor and its surface.

Types of Electrochemical Reactions

There are several types of electrochemical reactions:

Simple Redox Reactions

A simple redox reaction is one that involves a change in the electrical charge of a charge carrier, usually a simple or complex ion in the solution, by its taking away, an electron from the electrode (reduction), or its giving an electron to the electrode (oxidation). The same carrier may be present in solution in two states of charge. The higher, more positive charge is called the oxidized state, and the lower, less positive charge is called the reduced state. For example, when ferric and ferrous ions are both present in solution in significant quantity, and when electron exchange with the electrode is sufficiently fast, redox equilibrium is established at the electrode, giving it a well-defined potential, or reversible redox potential.

Reactions that Produce Gases

When hydrogen ions in solution react with electrons ejected from a metal, hydrogen atoms are formed at the surface, where they combine among themselves or with other hydrogen ions and electrons to give gaseous hydrogen molecules. If all the reactions are fast enough, an equilibrium is attained between hydrogen ions and gaseous hydrogen. A metal in contact with solution at which such a situation exists is called the reversible hydrogen electrode, and its electrical potential is arbitrarily taken to be zero; every other electrode can thus be compared with it as it represents the basis for constituting the hydrogen scale of relative electrode potentials. Similarly, negative hydroxyl ions in solution (OH^-) can be made to give up electrons to a metal and, in a series of reactions, the final one is the formation of gaseous oxygen. Chlorine is another gaseous product; it evolves upon electrochemical oxidation of chloride ions in concentrated solutions of neutral and acid salts.

Reactions that Deposit and Dissolve Metals

When a metal ion is reduced and discharged as a neutral atom, or species, it tends to build into the metal lattice of the electrode. Thus, metals can be deposited at electrodes. Conversely, if electrons are taken away from the metal electrode by applying positive potentials to it, the metal ions thus formed can cross the double layer of electric charge at the interface, undergo hydration (combination with water), and enter the solution. The metal electrode thus dissolves. Many metals establish well-defined electric potentials when they are in contact with their own ions in solution.

Oxidation and Reduction of Organic Compounds

A reaction of the oxidation and reduction of organic compounds can also be done at electrodes. Such reactions, however, are mostly irreversible in the literal sense that they lead to products that cannot easily be converted back into the original substance. Exceptions are some oxygen- and nitrogen-containing compounds (quinones, amines, and nitrous compounds) that can give fairly well-defined reversible potentials.

Mechanism of Charge Transfer

The causes of the thermodynamically irreversible behaviour of electrode reactions are found in the nature of the elementary act of charge transfer. Like any chemical reaction, this act is inhibited by the existence of an energy barrier between the oxidized and the reduced state. This barrier implies that the reaction could take place only in the special circumstances when, during the course of numerous interactions with other species (atoms, ions, molecules, etc.) surrounding it, a molecule attains an excited state in which it has an abnormal energy content. In most chemical reactions, this energy content must be sufficient for the species to come into what is called the transition state; the transition state characterizes the top of the energy barrier just before a reaction begins. If such a model is applied to electron transfer at an interface, calculation shows that electron exchange reactions at electrodes would be prohibitively slow, a conclusion at variance with the observed phenomena; quantum mechanical laws, however, govern the motion of electrons, and their inclusion changes the calculations to fit reality. Quantum mechanics require that for fast electron exchange to take place, electrons in a particle outside the double layer (e.g., a hydrated ion at the outer Helmholtz plane) must attain certain well-defined quantized energy levels equal to those in which free electrons exist in the metal. Since such states can be attained by the particle at a lower energy-content than that needed for its transfer over the top of the energy barrier, according to the classical view, this fast process of electron exchange between the electrode and a particle in solution is termed electron tunnelling through the energy barrier.

Whereas the rate of chemical processes, or what may loosely be termed the speed of reaction, can be influenced only by changing the concentrations of reactants or by changing the temperature or both, the rate of electrochemical processes also can be manipulated by changing the electrode potential. Making the electrode more negative increases the number of electrons in the metal ready to tunnel to ions, and hence the rate of the reduction process increases. Conversely, making the potential more positive decreases this rate and increases the number of particles ready to give away electrons, thus increasing the rate of the oxidation process.

It can be deduced that there must exist a direct proportionality between the rate of reaction and the concentration of the reacting species and at the same time an exponential proportionality between the rate of reaction and the electrode potential.

At any electrode potential, both reduction of one species and oxidation of the product of reduction are taking place but at different rates; the rate of each reaction is determined by the respective concentration and by the corresponding effects of potential. The rate of an electrochemical reaction can best be described as the electric current density—i.e., a measure of the quantity of electrons moving in a certain volume of space during a specified unit of time.

The relationships can be represented quantitatively by an equation in which the net, or resulting, current (the difference of the rate of electron ejection across the interface to particles in solution, diminished by the rate at which particles in solution inject electrons into the metal) is equated to the difference of the rates of reduction and oxidation and the variables and constants that relate to these reactions.

Complex Electrochemical Reactions

Electrochemical processes considered so far involve simple reactions of a particle with a single electron to produce a reduced ion (e.g., the ferrous ion of iron with two positive charges, Fe^{++}), or vice versa. Such are the simple ionic redox processes, where the only difference in structure between a reactant and a reaction product may be due to some rearrangement of the neighbouring solvent molecules. When one or more transfers of electrons between the electrode and a species in solution are accompanied by major structural changes (e.g., when hydroxyl ions, OH^-, transform into a molecule of oxygen, O_2, and a water molecule, H_2O, in the process of oxygen evolution at the anode, or positive electrode), the reaction usually consists of a sequence of events, called elementary acts, or unit steps, constituting the reaction mechanism. Intermediate states between the steps usually involve some unstable intermediate species with higher energy content than those of the reactants or of the reaction products.

Complex reaction mechanisms can consist of a number of electron transfer steps, with some chemical steps preceding or succeeding the electron transfer steps or taking place in between them. Most organic electrochemical reactions are complex, involving large numbers of electrons in the overall reaction. Usually one step in the reaction encounters the largest energy barrier. The rate of occurrence of this step limits the rate of the overall reaction (i.e., all other steps must occur at the same net rate, although they could provide for a much faster overall change). This step is called the rate-determining step and, for most practical purposes, all intermediate steps before and after it can be considered to be in equilibrium. (It is interesting to note that whenever this is the case, the Butler-Volmer equation is applicable, but with specific values of the transfer coefficients α_a and α_c characteristic of the mechanism of the reaction.)

Experimental Studies

Measuring the rates of electrochemical reactions (i.e., current densities) as functions of electrode potential under steady-state conditions represents the normal tool of electrodics. Meaningful results could not be obtained, however, until the sensitivity of electrochemical reactions to impurities was realized and high purity techniques were introduced. Even so, the steady-state method often has shortcomings except for relatively slow electrode reactions. In many cases concentration changes at the electrodes prevent using a sufficiently wide current density range for obtaining meaningful Tafel relationships. Hence, so- called transient methods have been developed in which one electrochemical factor in the situation is rationally perturbed and the time dependence of others observed. One such method consists of placing a constant current pulse upon an electrode and measuring the variation of the resulting current through the solution. This is called the galvanostatic method for measuring the rate of an electrochemical reaction. Applying a potential pulse while observing the variation of the rate as a function of time constitutes the potentiostatic method. A third method, called the potentiodynamic, or potential sweep, method involves observations of the current as a function of the potential, while the latter is varied at a constant, known rate.

The advantages of transient methods over steady-state ones, in which behaviour before the attainment of the steady-state is not part of the observation, are manifold. If observations are made at sufficiently short times, events can be recorded before the onset of concentration changes, and pure activation values can be found. Hence, Tafel relationships can be obtained over a larger current density range than if one makes measurements over longer times, as is required in the steady-state methods. The structure of the transient states can reveal important information, such as double layer capacitance and surface coverage of the electrode by intermediate species.

Several so-called kinetic parameters, for example, partial derivatives of current density and potential with respect to concentration of chosen reactants, can be extracted from experimental measurements.

Electrochemical measurements have a limited capacity to reveal the state of an electrode surface. Nonelectrochemical methods of studying electrode surfaces, therefore, have been stressed. Optical methods have considerably gained in importance. Ellipsometry (i.e., measuring changes in basic properties of polarized light as it is reflected from an electrode surface) was the first method that made possible a study of monomolecular layers of oxides and adsorbed oxygen as adsorbed organic molecules. Adaptation of such a method to transient use allows change in the surface to be related to the passivity of metals. Another kind of spectroscopy enables infrared spectra of species adsorbed at electrode surface to be taken. Mössbauer spectra may lead to an identification of thin layers on an electrode surface. A good future is seen for further development of refined techniques for the study of electrochemical processes by various combinations of spectroscopic and electrochemical means.

Types of Complex Reactions

Electrocatalysis

The problems related to the increase of rates of electrochemical reactions, or, to put it another way, the decrease of overpotential, needed to perform reactions at a given rate are the subject of electrocatalysis. Both increase and decrease are of considerable practical importance since they affect the economics of electrochemical processes. Electrocatalysis is concerned with the electrode as a substrate, or base, for electrochemical reaction and with the effect of its bulk and surface properties on the rate of reaction. Contrary to expectation, the rate of the basic step in electrochemical reaction—electron transfer—is independent of the ease with which electrons are released from the metal. Hence, those simple electrode processes in which no other changes but electron transfer take place have heats of activation virtually independent of the metal substrate.

In a reaction involving the formation of a chemical bond between the electrode substrate and one of the radicals (charged particles) formed on the surface, rates of reaction at a given potential may vary for different substrates by many orders of magnitude. The variation is a function of the strength (or energy) of bonds established between the intermediate species and the surface, and when the substrates are transition metals, regular relations between reaction rate and certain characteristics of the metal's electronic structure are observed. If such bonds are too weak, the catalytic effect of the substrate is small. If they are too strong, the intermediate species is too stable to react further, and the surface of the metal becomes virtually blocked for further reaction. Thus, there is an optimum bond strength that can be seen if the rate of a reaction is plotted against a property of the metal substrate, which is proportional to the ability of the surface to form bonds—e.g., its heat of sublimation.

Electrochemical reactions can also be catalyzed or inhibited by foreign species present in the electrolytic solution. In the most general case, ions can change the rates of reactions by changing the properties of the double layer by specific adsorption effects. Organic molecules inhibit electrochemical reaction by blocking the surface. For some organic ions or dipoles, the blocking effect can begin at a certain well-defined potential. Before attaining it, the electrode reaction gives the usual current density-potential relationship, which is then suddenly interrupted by an abnormal fall of the current, corresponding to adsorption.

Electrocrystallization

Deposition of metals and other substances at electrodes as a consequence of an electrode process exhibits a number of specific features. The electrode process is followed by crystal building, and this results in a continuous change of the electrode surface. This change, in turn, affects the electrochemical properties of the system—the double-layer capacity and the rate constants of the charge transfer processes. Hence, if electrochemical properties are to be studied, transient methods should be employed that allow measurements to be made before major changes in surface morphology (structure) take place.

Since the two steps, the discharge of ions at the electrode and the incorporation of the discharged ions into the crystal lattice, are separated in time and space, an intermediate species exists at the surface, that of relatively loosely bound and freely moving atoms, called adatoms. Since the electrons tend to join the rest in the bulk of the metal, adatoms appear to have a partial charge, less than that of the elementary positive charge. The adatoms therefore attract solvent molecules, and the species is partially solvated. This reaction justifies considering an adatom as a kind of adsorbed ion, called an adion, which, however, has already undergone partial discharge.

Because such an intermediate state is possible, ions need not be reduced to the neutral atomic state at the point of incorporation into the crystal lattice. Indeed, energy considerations reveal this reduction to be improbable. Instead, discharge of ions is favoured on crystal planes. Their motion from the location of discharge to a site on the crystal occurs by surface diffusion. When such surface diffusion is inhibited, it can, in some cases, control the rate of reaction.

Problems of the crystal-building step in electrocrystallization are in many respects identical with those of crystal growth from the gas phase. Crystal building can occur in any one of three possible ways.

Kinks, naturally existing at any metal surface, form a suitable half-lattice position at which an atom is surrounded by one-half of the number of atoms that would surround it in the bulk of the metal; there, adatoms can be successively trapped and thus the crystal lattice is extended along a crystal edge and further on across the surface. This step growth mechanism is shown in figure. The mechanism, however, has a limited capacity for crystal growth. A step can move as far as the edge of a crystal, and step growth would lead to smoothing of the surface to perfection, but then further growth would cease.

Mechanisms associated with screw dislocations, or twinning edges, can provide for a continuous growth of crystals. The screw dislocation mechanism, shown in figure, is made possible by a specific fault often found in the crystal lattice that may be called a dislocation originating from a shift of one atom in the lattice with respect to a perfect arrangement. This shift may then result in the formation of a monoatomic edge projecting above the electrode surface at which new atoms can be stacked. The stacking produces a turning of the edge around the base atom as a centre, the process producing a spiral growth.

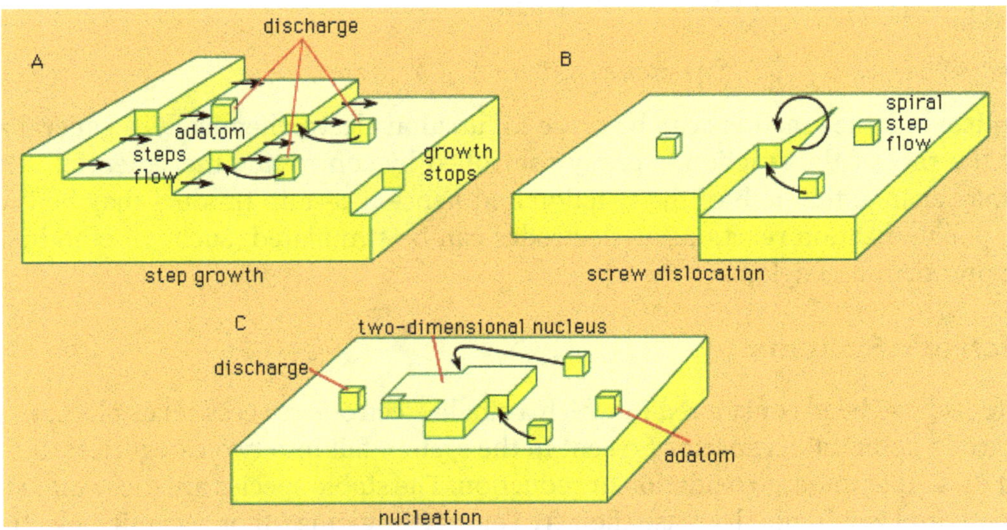

Mechanisms of crystal growth.

If growth sites are rare, or if the substrate at which deposition should take place is foreign to the depositing metal, the charge transfer results in an accumulation of adatoms to a concentration considerably larger than that which can exist there at equilibrium with the crystal lattice. In such a situation, termed supersaturation, agglomeration of adatoms to form crystal nuclei is favoured. Surface energy requirements show that, at any degree of supersaturation, nuclei of certain dimensions are stable and can represent sites for further growth, as shown in figure.

The formation of deposits is controlled by the above mechanism as long as the discharge process supplies ample amounts of adatoms as building material over the entire surface. If the rate of deposition is increased, so as to produce near the surface considerable depletion of ions, uneven deposition will start. This is caused by the protrusions of a normally rough metal surface being closer to the bulk of solution than the recessed parts and, hence, getting a somewhat faster supply of the discharging species. Once such a situation is established, it tends to develop further. The faster growing points penetrate into ever richer layers of solution, resulting in ever faster growth. Thus, a natural consequence of deposition under transport-controlled rate is the amplification of the original surface roughness and the appearance of protruding spikes, called dendrites, as deposits.

A converse process is the smoothing of the original surface irregularities, which may occur when some foreign species, called an additive, is present in solution and adsorbs on the surface and inhibits the process of discharge. If those molecules are incorporated into a growing deposit, a situation may arise in which their supply to recessed parts of the surface becomes slower than to elevated parts. As a result, deposition becomes faster at recessed parts than at elevated ones and leveling of the surface occurs. This process has considerable technical application. Metal dissolution can sometimes be governed by a similar, transport-controlled mechanism, in which case a polishing effect is obtained. Electropolishing has found wide application in practice.

Organic Eectrode Reactions

A very large number of electrochemical reactions involving organic molecules are known. An example is the oxidation of ethylene according to the equation:

$$\underset{(\text{ethylene})}{C_2H_4} + \underset{(\text{water})}{4H_2O} \rightarrow \underset{\left(\substack{\text{carbon}\\ \text{dioxide}}\right)}{2CO_2} + \underset{(\text{hydrogenion})}{12H^+} + \underset{(\text{elctrons})}{12e^-}.$$

Many chemical organic reactions can be made to function electrochemically, a general advantage being that the rate of the reaction is easily controlled by controlling the potential. A change of potential may change the path of the reaction and hence a certain product may be tuned in. In particular, polymerization reactions at electrodes can be stimulated; such a method is used as a step in the production of nylon.

Multielectrode Systems

So far, systems have been considered in which a single electrode process takes place. In principle, at any electrode potential all species present in the system fall into two categories: those that are stable, and those that undergo oxidation or reduction. The stable species are those that at the given electrode potential would not decrease their free energy by giving off or accepting electrons.

At any potential there should occur a codischarge of all unstable species. Thus, the system can be considered as a multielectrode system, consisting of as many electrodes as there are redox couples present. The rate at which different processes occur, however, can be so widely different that usually a single process is by far the dominant one. Systems at which two electrode processes occur at comparable rates are of considerable importance. If two kinds of metal ions are discharged simultaneously, an alloy is formed upon crystallization. The properties of the alloy would in most cases be those determined by the phase diagram (a plot of the temperature of melting versus the composition of the mixed system) for the ratio of quantities of discharged metals given by the rates at which the discharges take place. In many cases, nonequilibrium metal phases are formed giving unusual properties to the alloy.

If a process of metal dissolution (an oxidation process) can occur at a rate comparable to that of some reduction process on the same metal, a corrosion couple is established. Thus, zinc immersed in acid solution tends to establish a potential sufficiently positive for the metal to eject metal ions into the solution (oxidation). At the same time, it is also sufficiently negative for the reduction of hydrogen ions, present in any aqueous solution, to hydrogen gas. Hence, a spontaneous process of hydrogen evolution and of dissolution of the metal will take place. The mixed potential, spontaneously established by the corroding metal, is obtained by equating expressions for the anodic and cathodic currents of the two processes in the corrosion couple.

Corrosion can be prevented in two ways. One is by using an external source to make the potential of the metal sufficiently negative to bring it into the potential region in which the metal is stable, called cathodic protection; and the other is by provoking by some means the formation of a film on the surface that would slow the process. Such films could consist of an oxide or a layer of organic molecules that prevents dissolution and hence is called an inhibitor.

Applications

Electrochemical processes are used in many ways and their use is likely to increase because they can replace polluting chemical situations with nonpolluting electrochemical ones. In many fields, however, applications have been profitable for some time. Major categories are listed below.

Metallurgy

All technologically important metals, except iron and steel, are either obtained or refined by electrochemical processes; for example, aluminum, titanium, alkaline earth, and alkali metals are obtained by electrodeposition from molten salts, and copper is refined by electrolysis in aqueous copper sulfate solutions.

Electroplating

One of the major ways of both decorating objects and improving their resistance to corrosion is by electroplating them. All major metal-working industries, particularly the automobile industry, have large electroplating plants.

Chemical Industry

Electrolysis of brine to obtain chlorine and caustic soda is an electrochemical process that has become one of the largest volume productions in the chemical industry. Modern processes cover a wide field, from the production of a variety of inorganic compounds to the production of such synthetic fibres as nylon. Intensive research in organic electrochemistry promises major developments in application, particularly with the prospect of greatly reduced electricity costs expected eventually to arise from the development of controlled fusion.

Batteries

Electrochemical storage of electricity is effected in batteries. Such devices are electrochemical cells and consist of two electrodes per unit. As the electricity to be stored is accepted on the plates of the cell, it converts substances on the plates to new substances having a higher energy than the old ones. When it is desired to make the electricity available again, the terminals of the battery are connected to the load and the substances on the battery plates retransform themselves to those originally present, giving off electricity as a product of their electrochemical reactions. The steadily rising production of the lead-acid battery is largely the result of its use for starting the internal-combustion engine, which has had an equally steady rise. Other electrochemical systems are also used as storers. The nickel-iron (Edison cell) and nickel-cadmium battery with alkaline electrolyte are both used in applications where longer lives than those of the lead-acid battery are needed; the silver-zinc battery is used to start airplane engines because of its high power per unit of weight. A variety of new systems is being investigated for covering other needs. One of the greatest challenges to electrochemists and electrochemical engineers is that of producing a battery with sufficient power and energy density to run an automobile the way gasoline (petrol) does. Even if the best hypothetical predictions for removal of polluting chemicals from automobile exhausts is realized, the cleanup will not be sufficient because the expected growth of the automobile population will continue to increase the pollutant rate.

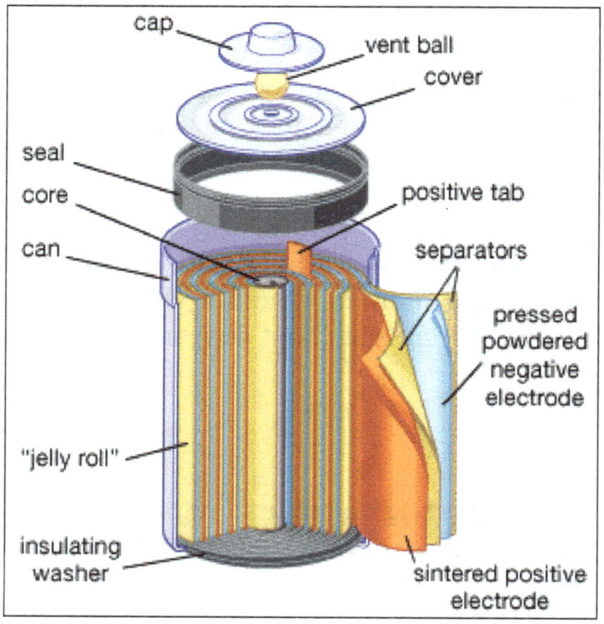

Nickel (hydroxide)-cadmium cell of "jelly roll" construction.
This rechargable battery is commonly used in portable devices.

Fuel Cells

The energy of chemical reactions is converted into electrical energy in fuel cells. In these, the fuel (e.g., hydrogen, hydrazine) is fed continuously to one electrode, while oxygen from the air is reacting at the other one. The efficiency of energy conversion in fuel cells is more than twice that attainable by conventional means—for example, by means of internal combustion.

Analytical Chemistry

In analytical chemistry, most modern automated instrumental analysis is based on electrode processes—for example, potentiometry, used to measure ionization constant.

Biological Research

In biology the idea that many biological processes, from blood clotting to the transfer of nerve impulses, are electrochemical in nature continues to spread. The biological conversion of the chemical energy of food to mechanical energy takes place at an efficiency so high that it is difficult to explain without electrochemical mechanisms. Intensive research is developing in various directions in bioelectrochemistry.

Electrochemical Cell

An electrochemical cell is a device that can generate electrical energy from the chemical reactions occurring in it, or use the electrical energy supplied to it to facilitate chemical reactions in it. These devices are capable of converting chemical energy into electrical energy, or vice versa. A common example of an electrochemical cell is a standard 1.5-volt cell which is used to power many electrical appliances such as TV remotes and clocks.

Such cells capable of generating an electric current from the chemical reactions occurring in them care called Galvanic cells or Voltaic cells. Alternatively, the cells which cause chemical reactions to occur in them when an electric current is passed through them are called electrolytic cells.

A diagram detailing the different parts of an electrochemical cell is provided below.

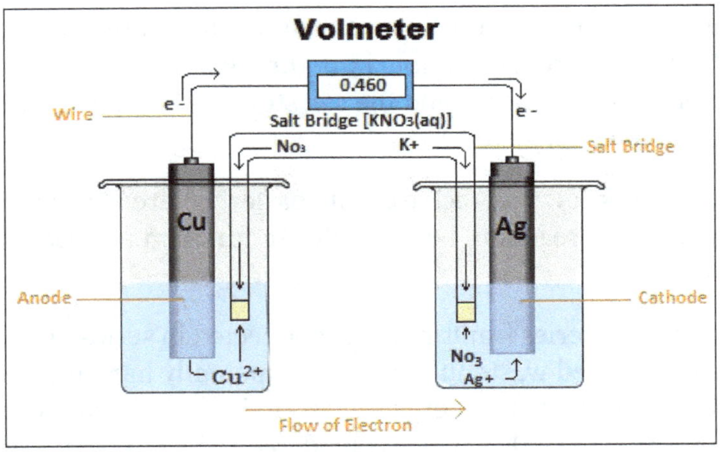

Electrochemical Cell.

Table: Electrochemical cells generally consist of a cathode and an anode. The key features of the cathode and the anode are tabulated below.

Cathode	Anode
Denoted by a positive sign since electrons are consumed here.	Denoted by a negative sign since electrons are liberated here.
A reduction reaction occurs in the cathode of an electrochemical cell.	An oxidation reaction occurs here.
Electrons move into the cathode.	Electrons move out of the anode.

General convention dictates that the cathode must be represented on the right-hand side whereas the anode is represented on the left-hand side while denoting an electrochemical cell.

Half-cells and Cell Potential

- Electrochemical Cells are made up of two half-cells, each consisting of an electrode which is dipped in an electrolyte. The same electrolyte can be used for both half cells.
- These half cells are connected by a salt bridge which provides the platform for ionic contact between them without allowing them to mix with each other. An example of a salt bridge is a filter paper which is dipped in a potassium nitrate or sodium chloride solution.
- One of the half cells of the electrochemical cell loses electrons due to oxidation and the other gains electrons in a reduction process. It can be noted that an equilibrium reaction occurs in both the half cells, and once the equilibrium is reached, the net voltage becomes 0 and the cell stops producing electricity.
- The tendency of an electrode which is in contact with an electrolyte to lose or gain electrons is described by its electrode potential. The values of these potentials can be used to predict the overall cell potential. Generally, the electrode potentials are measured with the help of the standard hydrogen electrode as a reference electrode (an electrode of known potential).

Primary and Secondary Cells

- Primary cells are basically use-and-throw galvanic cells. The electrochemical reactions that take place in these cells are irreversible in nature. Hence, the reactants are consumed for the generation of electrical energy and the cell stops producing an electric current once the reactants are completely depleted.
- Secondary cells (also known as rechargeable batteries) are electrochemical cells in which the cell has a reversible reaction, i.e. the cell can function as a Galvanic cell as well as an Electrolytic cell.
- Most of the primary batteries (multiple cells connected in series, parallel, or a combination of the two) are considered wasteful and environmentally harmful devices. This is because they require about 50 times the energy they contain in their manufacturing process. They also contain many toxic metals and are considered to be hazardous waste.

Galvanic Cell

Galvanic cells, also known as voltaic cells, are electrochemical cells in which spontaneous oxidation-reduction reactions produce electrical energy. In writing the equations, it is often convenient to separate the oxidation-reduction reactions into half-reactions to facilitate balancing the overall equation and to emphasize the actual chemical transformations.

Consider what happens when a clean piece of copper metal is placed in a solution of silver nitrate. As soon as the copper metal is added, silver metal begins to form and copper ions pass into the solution. The blue color of the solution on the far right indicates the presence of copper ions. The reaction may be split into its two half-reactions. Half-reactions separate the oxidation from the reduction, so each can be considered individually.

$$\text{oxidation: } Cu(s) \rightarrow Cu^{2+}(aq) + 2e^-$$

$$\text{reduction: } 2 \times \left(Ag^+(aq) + e^- \rightarrow Ag(s)\right) \; or \; 2Ag^+(aq) + 2e^- \rightarrow Ag(s)$$

$$\text{overall: } 2Ag^+(aq) + Cu(s) \rightarrow 2g(s) + Cu^{2+}(aq)$$

The equation for the reduction half-reaction had to be doubled so the number electrons "gained" in the reduction half-reaction equaled the number of electrons "lost" in the oxidation half-reaction.

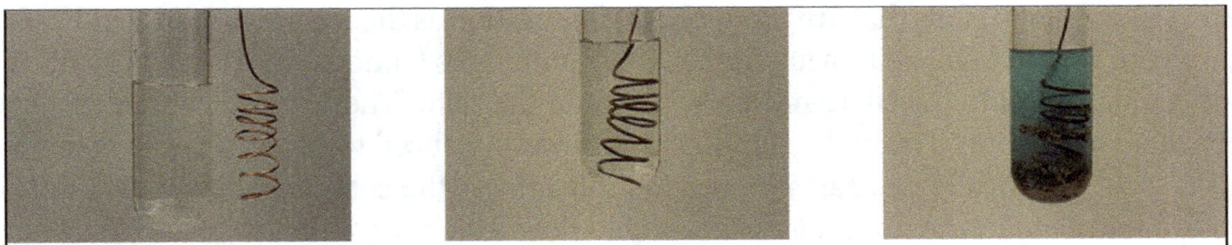
(a) (b) (c)

When a clean piece of copper metal is placed into a clear solution of silver nitrate (a), an oxidation-reduction reaction occurs that results in the exchange of Cu^{2+} for Ag^+ ions in solution. As the reaction proceeds (b), the solution turns blue (c) because of the copper ions present, and silver metal is deposited on the copper strip as the silver ions are removed from solution.

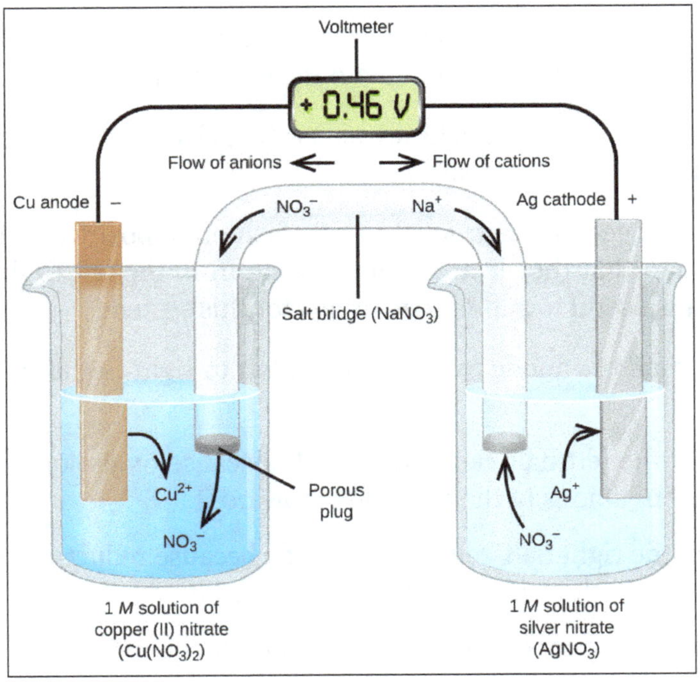

Galvanic or voltaic cells involve spontaneous electrochemical reactions in which the half-reactions are separated so that current can flow through an external wire. The beaker on the left side of the figure is called a half-cell, and contains a 1 *M* solution of copper(II) nitrate [$Cu(NO_3)_2$] with a piece of copper metal partially submerged in the solution. The copper metal is an electrode. The copper is undergoing oxidation; therefore, the copper electrode is the anode. The anode is connected to a voltmeter with a wire and the other terminal of the voltmeter is connected to a silver electrode by a wire. The silver is undergoing reduction; therefore, the silver electrode is the cathode. The half-cell on the right side of the figure consists of the silver electrode in a 1 *M* solution of silver nitrate ($AgNO_3$). At this point, no current flows—that is, no significant movement of electrons through the wire occurs because the circuit is open. The circuit is closed using a salt bridge, which transmits the current with moving ions. The salt bridge consists of a concentrated, nonreactive, electrolyte solution such as the sodium nitrate ($NaNO_3$) solution used in this example. As electrons flow from left to right through the electrode and wire, nitrate ions (anions) pass through the porous plug on the left into the copper(II) nitrate solution. This keeps the beaker on the left electrically neutral by neutralizing the charge on the copper(II) ions that are produced in the solution as the copper metal is oxidized. At the same time, the nitrate ions are moving to the left, sodium ions (cations) move to the right, through the porous plug, and into the silver nitrate solution on the right. These added cations "replace" the silver ions that are removed from the solution as they were reduced to silver metal, keeping the beaker on the right electrically neutral. Without the salt bridge, the compartments would not remain electrically neutral and no significant current would flow. However, if the two compartments are in direct contact, a salt bridge is not necessary. The instant the circuit is completed, the voltmeter reads +0.46 V, this is called the cell potential. The cell potential is created when the two dissimilar metals are connected, and is a measure of the energy per unit charge available from the oxidation-reduction reaction. The volt is the derived SI unit for electrical potential:

$$\text{volt} = \text{V} = \frac{1}{\text{C}}$$

In this equation, A is the current in amperes and C the charge in coulombs. Note that volts must be multiplied by the charge in coulombs (C) to obtain the energy in joules (J).

In this standard galvanic cell, the half-cells are separated; electrons can flow through an external wire and become available to do electrical work.

When the electrochemical cell is constructed in this fashion, a positive cell potential indicates a spontaneous reaction and that the electrons are flowing from the left to the right. There is a lot going on in figure, so it is useful to summarize things for this system:

- Electrons flow from the anode to the cathode: left to right in the standard galvanic cell in the figure.
- The electrode in the left half-cell is the anode because oxidation occurs here. The name refers to the flow of anions in the salt bridge toward it.
- The electrode in the right half-cell is the cathode because reduction occurs here. The name refers to the flow of cations in the salt bridge toward it.
- Oxidation occurs at the anode (the left half-cell in the figure).
- Reduction occurs at the cathode (the right half-cell in the figure).
- The cell potential, +0.46 V, in this case, results from the inherent differences in the nature of the materials used to make the two half-cells.
- The salt bridge must be present to close (complete) the circuit and both an oxidation and reduction must occur for current to flow.

There are many possible galvanic cells, so a shorthand notation is usually used to describe them. The cell notation (sometimes called a cell diagram) provides information about the various species involved in the reaction. This notation also works for other types of cells. A vertical line, |, denotes a phase boundary and a double line, ||, the salt bridge. Information about the anode is written to the left, followed by the anode solution, then the salt bridge (when present), then the cathode solution, and, finally, information about the cathode to the right. The cell notation for the galvanic cell in figure is then:

$$\mathrm{Cu}(s)\mathrm{Cu}^{2+}(aq,1M)\|\mathrm{Ag}^{+}(aq,1M)|\mathrm{Ag}(s)$$

Note that spectator ions are not included and that the simplest form of each half-reaction was used. When known, the initial concentrations of the various ions are usually included.

One of the simplest cells is the Daniell cell. It is possible to construct this battery by placing a copper electrode at the bottom of a jar and covering the metal with a copper sulfate solution. A zinc sulfate solution is floated on top of the copper sulfate solution; then a zinc electrode is placed in the zinc sulfate solution. Connecting the copper electrode to the zinc electrode allows an electric current to flow. This is an example of a cell without a salt bridge, and ions may flow across the interface between the two solutions.

Some oxidation-reduction reactions involve species that are poor conductors of electricity, and so an electrode is used that does not participate in the reactions. Frequently, the electrode is platinum, gold, or graphite, all of which are inert to many chemical reactions. One such system is shown in figure. Magnesium undergoes oxidation at the anode on the left in the figure and hydrogen ions undergo reduction at the cathode on the right. The reaction may be summarized as:

$$\text{oxidation:}\quad \mathrm{Mg(s)}\rightarrow\mathrm{Mg}^{2+}(aq)+2e^{-}$$

$$\text{reduction:}\quad 2\mathrm{H}^{+}(aq)+2e^{-}\rightarrow\mathrm{H}_2(g)$$

$$\text{overall:}\quad \mathrm{Mg}(s)+2\mathrm{H}^{+}(aq)\rightarrow\mathrm{Mg}^{2+}(aq)+\mathrm{H}_2(g)$$

The cell used an inert platinum wire for the cathode, so the cell notation is:

$$\mathrm{Mg}(s)\mathrm{Mg}^{2+}(aq.)\|\mathrm{H}^{+}(aq)\mathrm{H}_2(g)|\mathrm{Pt}(s)$$

The magnesium electrode is an active electrode because it participates in the oxidation-reduction reaction. Inert electrodes, like the platinum electrode in figure, do not participate in the oxidation-reduction reaction and are present so that current can flow through the cell. Platinum or gold generally make good inert electrodes because they are chemically unreactive.

Example:

Using Cell Notation

Consider a galvanic cell consisting of:

$$2\mathrm{Cr}(s)+3\mathrm{Cu}^{2+}(aq)\rightarrow2\mathrm{Cr}^{3+}(aq)+3\mathrm{Cu}(s)$$

Write the oxidation and reduction half-reactions and write the reaction using cell notation. Which reaction occurs at the anode? The cathode?

Solution: By inspection, Cr is oxidized when three electrons are lost to form Cr^{3+}, and Cu^{2+} is reduced as it gains two electrons to form Cu. Balancing the charge gives:

$$\text{oxidation:} \quad 2\text{Cr}(s) \rightarrow 2\text{Cr}^{3+}(aq) + 6e^-$$

$$\text{reduction:} \quad 3\text{Cu}^{2+}(aq) + 6e^- \rightarrow 3\text{Cu}(s)$$

$$\text{overall:} \quad 2\text{Cr}(s) + 3\text{Cu}^{2+}(aq) \rightarrow 2\text{Cr}^{3+}(aq) + 3\text{Cu}(s)$$

Cell notation uses the simplest form of each of the equations, and starts with the reaction at the anode. No concentrations were specified so: $\text{Cr}(s)|\text{Cr}^{3+}(aq)\,\|\text{Cu}^{2+}(aq)\,|\text{Cu}(s)$. Oxidation occurs at the anode and reduction at the cathode.

Consider a galvanic cell consisting of;

$$5\text{Fe}^{2+}(aq) + \text{MnO}_4^-(aq) + 8\text{H}^+(aq) \rightarrow 5\text{Fe}^{3+}(aq) + \text{Mn}^{2+}(aq) + 4\text{H}_2\text{O}(l)$$

Write the oxidation and reduction half-reactions and write the reaction using cell notation. Which reaction occurs at the anode? The cathode?

Solution: By inspection, Fe^{2+} undergoes oxidation when one electron is lost to form Fe^{3+}, and MnO_4^- is reduced as it gains five electrons to form Mn^{2+}. Balancing the charge gives:

$$\text{oxidation:} \quad 5\left(\text{Fe}^{2+}(aq) \rightarrow 5\text{Fe}^{3+}(aq) + e^-\right)$$

$$\text{reduction:} \quad \text{MnO}_4^-(aq) + 8\text{H}^+(aq) + 5e^- + \text{Mn}^{2+}(aq) \rightarrow 4\text{H}_2\text{O}(l)$$

$$\text{overall:} \quad 5\text{Fe}^{2+}(aq) + \text{MnO}_4^-(aq) + 8\text{H}^+(aq) \rightarrow 5\text{Fe}^{3+}(aq)\text{Mn}^{2+}(aq) + 4\text{H}_2\text{O}(l)$$

Cell notation uses the simplest form of each of the equations, and starts with the reaction at the anode. It is necessary to use an inert electrode, such as platinum, because there is no metal present to conduct the electrons from the anode to the cathode. No concentrations were specified so: $\text{Pt}(s)|5\text{Fe}^{2+}(aq) + 5\text{Fe}^{3+}(aq)\|\text{MnO}_4^-(aq)\text{H}^+(aq) + \text{Mn}^{2+}(aq)|\,\text{Pt}(s)$. Oxidation occurs at the anode and reduction at the cathode.

Example: Use cell notation to describe the galvanic cell where copper(II) ions are reduced to copper metal and zinc metal is oxidized to zinc ions.

Answer:

From the information given in the problem:

$$\text{anode(oxidation):} \quad \text{Zn}(s) \rightarrow \text{Zn}^{2+}(aq) + 2e^-$$

$$\text{cathode(reduction):} \quad \text{Cu}^{2+}(aq) + 2\text{e}^- \rightarrow \text{Cu}(s)$$

$$\text{overall:} \quad \text{Zn}(s) + \text{Cu}^{2+}(aq) \rightarrow \text{Zn}^{2+}(aq) + \text{Cu}(s)$$

Using cell notation:

$$Zn(s)|Zn^{2+}(aq)||Cu^{2+}(aq)|Cu(s)$$

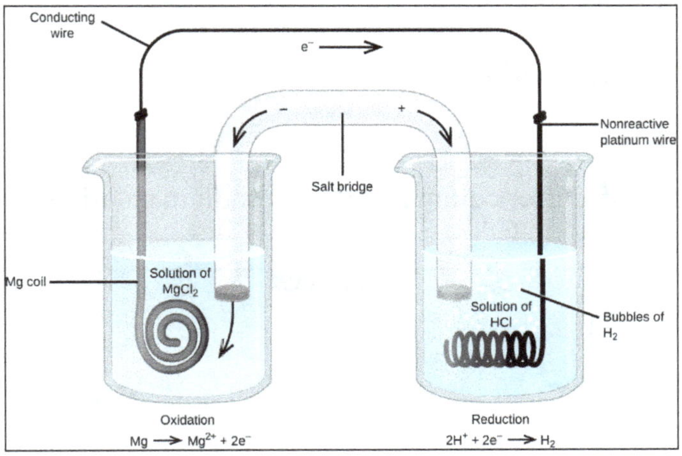

The oxidation of magnesium to magnesium ion occurs in the beaker on the left side in this apparatus; the reduction of hydrogen ions to hydrogen occurs in the beaker on the right. A nonreactive, or inert, platinum wire allows electrons from the left beaker to move into the right beaker. The overall reaction is: $Mg + 2H^+ \longrightarrow Mg^{2+} + H_2$, which is represented in cell notation as: $Mg(s) \mid Mg^{2+}(aq) \mid\mid H^+(aq) \mid H_2(g) \mid Pt(s)$.

Electrochemical cells typically consist of two half-cells. The half-cells separate the oxidation half-reaction from the reduction half-reaction and make it possible for current to flow through an external wire. One half-cell, normally depicted on the left side in a figure, contains the anode. Oxidation occurs at the anode. The anode is connected to the cathode in the other half-cell, often shown on the right side in a figure. Reduction occurs at the cathode. Adding a salt bridge completes the circuit allowing current to flow. Anions in the salt bridge flow toward the anode and cations in the salt bridge flow toward the cathode. The movement of these ions completes the circuit and keeps each half-cell electrically neutral. Electrochemical cells can be described using cell notation. In this notation, information about the reaction at the anode appears on the left and information about the reaction at the cathode on the right. The salt bridge is represented by a double line, ||. The solid, liquid, or aqueous phases within a half-cell are separated by a single line, |. The phase and concentration of the various species is included after the species name. Electrodes that participate in the oxidation-reduction reaction are called active electrodes. Electrodes that do not participate in the oxidation-reduction reaction but are there to allow current to flow are inert electrodes. Inert electrodes are often made from platinum or gold, which are unchanged by many chemical reactions.

Electrolytic Cell

Electrolytic cells convert electrical energy into chemical potential energy. The process is known as electrolysis. The purpose of this is usually to convert reactants into more useful products.

Electrolytic cells are one of two major categories of electrochemical cell.

The other category, voltaic cells, convert chemical potential energy to electrical energy. Battery-powered devices get their electrical energy from one or more voltaic cells.

Thermodynamics

Electrolytic cells use electrical energy to drive non-spontaneous chemical reactions - i.e. those reactions for which $\Delta G \geq 0$.

Examples of Electrolytic Cells

Manufacturing Sodium

Electrolysis is used to manufacture sodium metal from sodium chloride.

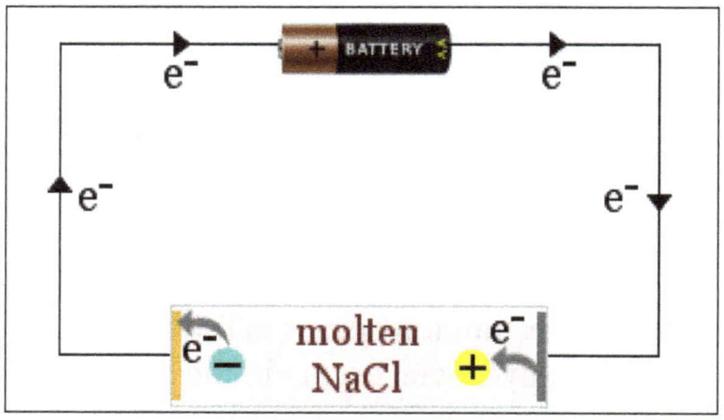

$$Cl^-_{(l)} \rightarrow \tfrac{1}{2}Cl_{2(g)} + e^- \quad Na^+_{(l)} + e^- \rightarrow Na_{(l)}$$

Recharging Batteries

Electrolysis is used to recharge rechargeable batteries: rechargeable batteries operate as voltaic cells when they are powering devices and as electrolytic cells during recharge.

For example, the Edison battery is a simple, rechargeable cell invented by Thomas Edison. It consists of two metal electrodes, one made of iron, the other of nickel. During initial charging, a coating of nickel oxide forms on the nickel electrode.

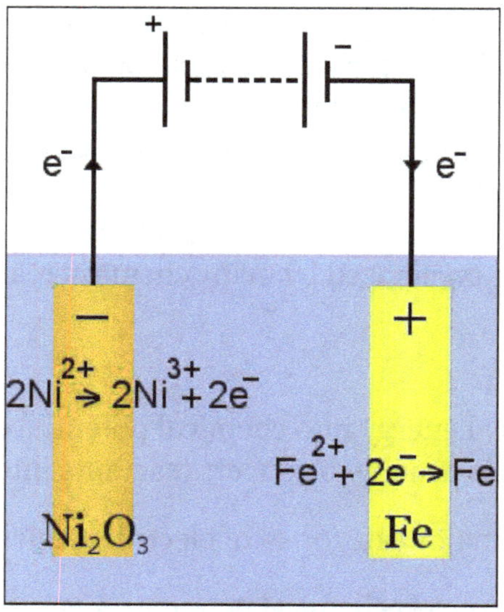

Crude schematic of Edison cell charging.

The electrolyte (the ionic liquid between the electrodes) is aqueous potassium hydroxide.

When it is discharging, the Edison cell operates as a voltaic cell. When it is being charged, the cell operates as an electrolytic cell.

The chemical equations for the reactions at the electrodes are:

$$Ni_2O_3 + H_2O + 2\ e^- \rightleftharpoons 2\ NiO + 2\ OH^-$$

$$Fe + 2\ OH^- \rightleftharpoons Fe(OH)_2 + 2\ e^-$$

During discharge, when the cell is delivering electrical energy, the reactions above proceed from left to right.

During charging, when the cell is operating electrolytically, converting electrical energy to chemical potential energy, the reactions above proceed from right to left.

Electrolysis of Water

Electrolysis of water produces the ultimate in clean fuels; one of chemistry's Holy Grails is to split water using sunlight.

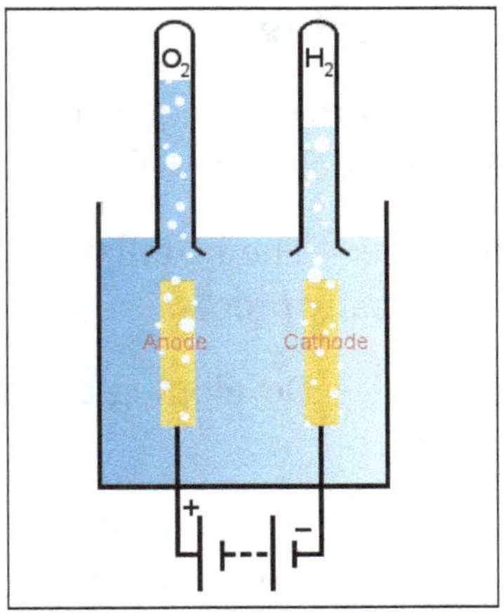

Direct current can be used to split water.

At the cathode, hydrogen ions gain electrons:

$$2H^+(aq) + 2e^- \rightarrow H_2(g)$$

At the anode, water loses electrons, forming oxygen and hydrogen ions:

$$2H_2O(l) \rightarrow O_2(g) + 4H^+(aq) + 4e^-$$

Process Engineering

Process Engineering is a study of engineering that focuses on the design and optimization of biological and chemical processes.

Process engineering is the means by which raw materials are turned into some kind of end product or commodity. Process engineers are responsible for designing the processes that allow for the creation of some kind of product – as well as their optimisation to ensure that there is minimal waste during the process, and profits are maximised.

To this end, process engineering is a very complex engineering specialisation. Process engineers must create and develop product specifications, develop equipment requirements, implement manufacturing strategies, and monitor all of the processes they develop – to ensure maximum productivity.

These processes vary from simple mechanical processes – such as the stamping of metal or CNC machining of plastic components – to complex biochemical reactions, and everything in-between. While process engineers do develop new, innovative processes for new facilities, they are also often involved in the analysis, upgrading, and modification/optimisation of equipment and processes that are already being used by a company.

Process engineering is an inherently cross-departmental discipline. Process engineers usually must work with people from every sector of the manufacturing industry – including R&D staff, production personnel, operations personnel, management, and even customers, in some cases.

Areas of Focus in Process Engineering

There are hundreds of specialisations that process engineers can choose. However, there are some common areas of focus that many engineers choose to focus on during their careers:

- Asset Management & Maintenance – The installation, management, and maintenance of new equipment is a key role that many process engineers choose to focus on.
- Process Engineering/Management – Developing and managing new processes is a common role for all process engineers.
- Process Automation/Control – Controlling and automating processes requires a specialised skillset, usually including some kind of programming or computer science background.
- Drives & Motors – Optimising drives, motors, and other industrial equipment is a specialised area of focus, often taken on by process engineers with a mechanical engineering background.
- Process Instrumentation & Measurement – Developing instrumentation and measurement techniques is a key part of any process engineering project.
- Fluids/Liquids Handling And Processing – Liquids and fluids handling expertise is usually required in the food & beverage industry, as well as other industries like utilities, wastewater management, and petrochemical manufacturing.
- Solids Handling/Processing – Solids handling is a large field, encompassing areas from food service to paper processing and metallurgy/mining.
- Health & Safety Analysis – Health and safety analysts focus on the safety of personnel and equipment, as well as the streamlining of common manufacturing processes.

- Industrial Networking & Communication – Communication and networking in industrial settings is essential for proper machine automation and process performance. Process engineers with an IT background are highly sought-after for these positions.
- Industrial Internet of Things (IIoT) – The Industrial Internet of Things (IIoT) utilises modern IoT technology to provide manufacturers with more in-depth analyses of their process performance. Engineers who focus on this area often combine IT and programming expertise with deep knowledge of a particular manufacturing industry.

Process Engineering Industries

Process engineering is valuable for any industry that manufactures consumer products, processes raw materials, or otherwise must utilise advanced production technologies to create some kind of saleable product. Here are a few industries where process engineering is highly-valued:

- Food & Beverage – Food and beverage manufacturers need process engineers to ensure that their processes are efficient, food-safe, and maximise profitability.
- Chemicals – Chemical process engineering is highly complex, requiring an in-depth knowledge of both chemistry and process engineering.
- Paper & Pulp – Paper mills and pulping plants often utilise process engineers to ensure maximum performance of all existing processes, and to create novel and innovative manufacturing strategies.
- Plastics & Rubber – As a subset of the petrochemical industry, plastics & rubber companies must use process engineers to ensure product quality, minimal waste, and adherence to environmental regulations.
- Metals & Minerals – Mining and material extraction companies in the metals/minerals industry use process engineers to ensure minimal waste of valuable products during the refining process.
- Ceramics, Glass & Cement – Because ceramics, glass, and cement are often proprietary and use distinct manufacturing processes, process engineers are highly sought-after to improve each process, and maximise profitability.
- Utilities – Process engineering is extremely valuable to utility companies. This is because utility companies often have a difficult time increasing profits – and streamlining delivery and minimising waste is one of the few ways to guarantee more profitability.
- Power Generation – Process engineers are required for all power plants and power generation centers; from coal-fired power plants, to natural gas processing facilities, nuclear power plants, and more.
- Water & Wastewater – Processing water and wastewater usually falls to municipal authorities, who use process engineers to ensure maximum efficiency and safety of their processing plants.

- Pharmaceuticals – Process engineering is key for pharmaceutical companies. Each individual drug must be manufactured efficiently, but with razor-thin tolerances for chemical composition and effectiveness, making efficient and safe processes absolutely critical.

References

- Bioreactors-technology-design-analysis: researchgate.net, Retrieved 31 March, 2019
- Electrochemical-reaction, science: britannica.com, Retrieved 14 July, 2019
- Electrochemical-cell, chemistry: byjus.com, Retrieved 17 May, 2019
- 17-2-galvanic-cells, chemistry: opentextbc.ca, Retrieved 5 February, 2019
- Electrolytic-cell, definition: chemicool.com, Retrieved 26 July, 2019
- Process-engineering-everything-need-know, editorial: processindustryinformer.com, Retrieved 21 May, 2019

3

Chemical Reaction Engineering

The specialty within chemical engineering which focuses on studying and optimizing chemical reactions for the purpose of defining the best reactor design is known as chemical reaction engineering. All the diverse areas of study related to chemical reaction engineering such as chemical reactions, chemical kinetics and chemical reactors have been carefully analyzed in this chapter.

Chemical Reaction

Chemical reaction is a process in which one or more substances, the reactants, are converted to one or more different substances, the products. Substances are either chemical elements or compounds. A chemical reaction rearranges the constituent atoms of the reactants to create different substances as products.

Chemical reactions are an integral part of technology, of culture, and indeed of life itself. Burning fuels, smelting iron, making glass and pottery, brewing beer, and making wine and cheese are among many examples of activities incorporating chemical reactions that have been known and used for thousands of years.

Chemical reactions must be distinguished from physical changes. Physical changes include changes of state, such as ice melting to water and water evaporating to vapour. If a physical change occurs, the physical properties of a substance will change, but its chemical identity will remain the same. No matter what its physical state, water (H_2O) is the same compound, with each molecule composed of two atoms of hydrogen and one atom of oxygen. However, if water, as ice, liquid, or vapour, encounters sodium metal (Na), the atoms will be redistributed to give the new substances molecular hydrogen (H_2) and sodium hydroxide (NaOH). By this, we know that a chemical change or reaction has occurred.

Basic Concepts of Chemical Reactions

Synthesis

When making a new substance from other substances, chemists say either that they carry out a synthesis or that they synthesize the new material. Reactants are converted to products, and the process is symbolized by a chemical equation. For example, iron (Fe) and sulfur (S) combine to form iron sulfide (FeS).

$$Fe(s) + S(s) \rightarrow FeS(s)$$

The plus sign indicates that iron reacts with sulfur. The arrow signifies that the reaction "forms" or "yields" iron sulfide, the product. The state of matter of reactants and products is designated with the symbols (s) for solids, (l) for liquids, and (g) for gases.

Conservation of Matter

In reactions under normal laboratory conditions, matter is neither created nor destroyed, and elements are not transformed into other elements. Therefore, equations depicting reactions must be balanced; that is, the same number of atoms of each kind must appear on opposite sides of the equation. The balanced equation for the iron-sulfur reaction shows that one iron atom can react with one sulfur atom to give one formula unit of iron sulfide.

Chemists ordinarily work with weighable quantities of elements and compounds. For example, in the iron-sulfur equation the symbol Fe represents 55.845 grams of iron, S represents 32.066 grams of sulfur, and FeS represents 87.911 grams of iron sulfide. Because matter is not created or destroyed in a chemical reaction, the total mass of reactants is the same as the total mass of products. If some other amount of iron is used, say, one-tenth as much (5.585 grams), only one-tenth as much sulfur can be consumed (3.207 grams), and only one-tenth as much iron sulfide is produced (8.791 grams). If 32.066 grams of sulfur were initially present with 5.585 grams of iron, then 28.859 grams of sulfur would be left over when the reaction was complete.

The reaction of methane (CH_4, a major component of natural gas) with molecular oxygen (O_2) to produce carbon dioxide (CO_2) and water can be depicted by the chemical equation:

$$CH_4(g) + 2O_2(g) \rightarrow CO_2(g) + 2H_2O(l)$$

Here another feature of chemical equations appears. The number 2 preceding O_2 and H_2O is a stoichiometric factor. (The number 1 preceding CH_4 and CO_2 is implied.) This indicates that one molecule of methane reacts with two molecules of oxygen to produce one molecule of carbon dioxide and two molecules of water. The equation is balanced because the same number of atoms of each element appears on both sides of the equation (here one carbon, four hydrogen, and four oxygen atoms). Analogously with the iron-sulfur example, we can say that 16 grams of methane and 64 grams of oxygen will produce 44 grams of carbon dioxide and 36 grams of water. That is, 80 grams of reactants will lead to 80 grams of products.

The ratio of reactants and products in a chemical reaction is called chemical stoichiometry. Stoichiometry depends on the fact that matter is conserved in chemical processes, and calculations giving mass relationships are based on the concept of the mole. One mole of any element or compound contains the same number of atoms or molecules, respectively, as one mole of any other element or compound. By international agreement, one mole of the most common isotope of carbon (carbon-12) has a mass of exactly 12 grams (this is called the molar mass) and represents $6.022140857 \times 10^{23}$ atoms (Avogadro's number). One mole of iron contains 55.847 grams; one mole of methane contains 16.043 grams; one mole of molecular oxygen is equivalent to 31.999 grams; and one mole of water is 18.015 grams. Each of these masses represents $6.022140857 \times 10^{23}$ molecules.

Energy Considerations

Energy plays a key role in chemical processes. According to the modern view of chemical reactions, bonds between atoms in the reactants must be broken, and the atoms or pieces of molecules are reassembled into products by forming new bonds. Energy is absorbed to break bonds, and energy is evolved as bonds are made. In some reactions the energy required to break bonds is larger than the energy evolved on making new bonds, and the net result is the absorption of energy. Such a reaction is said to be endothermic if the energy is in the form of heat. The opposite of endothermic is exothermic; in an exothermic reaction, energy as heat is evolved. The more general terms exoergic (energy evolved) and endoergic (energy required) are used when forms of energy other than heat are involved.

A great many common reactions are exothermic. The formation of compounds from the constituent elements is almost always exothermic. Formation of water from molecular hydrogen and oxygen and the formation of a metal oxide such as calcium oxide (CaO) from calcium metal and oxygen gas are examples. Among widely recognizable exothermic reactions is the combustion of fuels.

The formation of slaked lime (calcium hydroxide, $Ca(OH)_2$) when water is added to lime (CaO) is exothermic:

$$CaO(s) + H2O\ (l) \rightarrow Ca(OH)_2(s)$$

This reaction occurs when water is added to dry portland cement to make concrete, and heat evolution of energy as heat is evident because the mixture becomes warm.

Not all reactions are exothermic (or exoergic). A few compounds, such as nitric oxide (NO) and hydrazine (N_2H_4), require energy input when they are formed from the elements. The decomposition of limestone ($CaCO_3$) to make lime (CaO) is also an endothermic process; it is necessary to heat limestone to a high temperature for this reaction to occur:

$$CaCO_3(s) \rightarrow CaO(s) + CO_2(g)$$

The decomposition of water into its elements by the process of electrolysis is another endoergic process. Electrical energy is used rather than heat energy to carry out this reaction:

$$2\ H_2O(g) \rightarrow 2\ H_2(g) + O_2(g)$$

Generally, evolution of heat in a reaction favours the conversion of reactants to products. However, entropy is important in determining the favourability of a reaction. Entropy is a measure of the number of ways in which energy can be distributed in any system. Entropy accounts for the fact that not all energy available in a process can be manipulated to do work.

A chemical reaction will favour the formation of products if the sum of the changes in entropy for the reaction system and its surroundings is positive. An example is burning wood. Wood has a low entropy. When wood burns, it produces ash as well as the high-entropy substances carbon dioxide gas and water vapour. The entropy of the reacting system increases during combustion. Just as important, the heat energy transferred by the combustion to its surroundings increases the entropy in the surroundings. The total of entropy changes for the substances in the reaction and the surroundings is positive, and the reaction is product-favoured.

When hydrogen and oxygen react to form water, the entropy of the products is less than that of the reactants. Offsetting this decrease in entropy, however, is the increase in entropy of the surroundings owing to the heat transferred to it by the exothermic reaction. Again because of the overall increase in entropy, the combustion of hydrogen is product-favoured.

Kinetic Considerations

Chemical reactions commonly need an initial input of energy to begin the process. Although the combustion of wood, paper, or methane is an exothermic process, a burning match or a spark is needed to initiate this reaction. The energy supplied by a match arises from an exothermic chemical reaction that is itself initiated by the frictional heat generated by rubbing the match on a suitable surface.

In some reactions, the energy to initiate a reaction can be provided by light. Numerous reactions in Earth's atmosphere are photochemical, or light-driven, reactions initiated by solar radiation. One example is the transformation of ozone (O_3) into oxygen (O_2) in the troposphere. The absorption of ultraviolet light (*hv*) from the Sun to initiate this reaction prevents potentially harmful high-energy radiation from reaching Earth's surface.

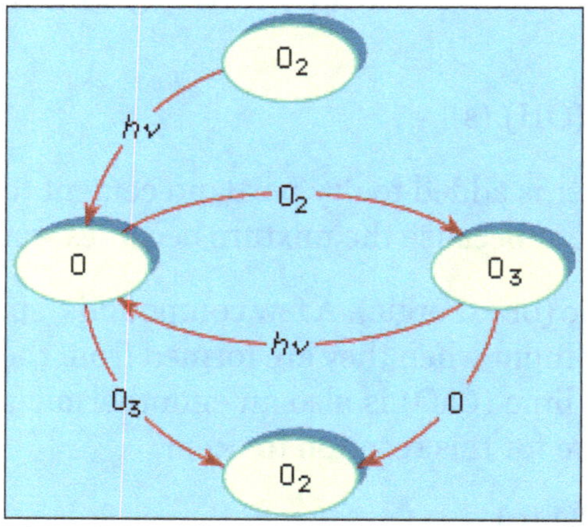

Schematic view of ozone chemistry in a pure oxygen environment.
Ultraviolet light is represented by *hv*.

For a reaction to occur, it is not sufficient that it be energetically product-favoured. The reaction must also occur at an observable rate. Several factors influence reaction rates, including the concentrations of reactants, the temperature, and the presence of catalysts. The concentration affects the rate at which reacting molecules collide, a prerequisite for any reaction. Temperature is influential because reactions occur only if collisions between reactant molecules are sufficiently energetic. The proportion of molecules with sufficient energy to react is related to the temperature. Catalysts affect rates by providing a lower energy pathway by which a reaction can occur. Among common catalysts are precious metal compounds used in automotive exhaust systems that accelerate the breakdown of pollutants such as nitrogen dioxide into harmless nitrogen and oxygen. A wide array of biochemical catalysts are also known, including chlorophyll in plants (which facilitates the reaction by which atmospheric carbon dioxide is converted to complex organic molecules such as glucose) and many biochemical catalysts called enzymes. The enzyme pepsin, for example, assists in the breakup of large protein molecules during digestion.

Classifying Chemical Reactions

Chemists classify reactions in a number of ways: (a) by the type of product, (b) by the types of reactants, (c) by reaction outcome, and (d) by reaction mechanism. Often, a given reaction can be placed in two or even three categories.

Classification by Type of Product

Gas-forming Reactions

Many reactions produce a gas such as carbon dioxide, hydrogen sulfide (H_2S), ammonia (NH_3), or sulfur dioxide (SO_2). An example of a gas-forming reaction is that which occurs when a metal carbonate such as calcium carbonate ($CaCO_3$, the chief component of limestone, seashells, and marble) is mixed with hydrochloric acid (HCl) to produce carbon dioxide:

$$CaCO_3(s) + 2\ HCl(aq) \rightarrow CaCl_2(aq) + CO_2(g) + H_2O\ (l)$$

In this equation, the symbol (aq) signifies that a compound is in an aqueous, or water, solution.

Cake-batter rising is caused by a gas-forming reaction between an acid and baking soda, sodium hydrogen carbonate (sodium bicarbonate, $NaHCO_3$). Tartaric acid ($C_4H_6O_6$), an acid found in many foods, is often the acidic reactant:

$$C_4H_6O_6(aq) + NaHCO_3(aq) \rightarrow NaC_4H_5O_6(aq) + H_2O\ (l) + CO_2(g)$$

In this equation, $NaC_4H_5O_6$ is sodium tartrate.

Bread-dough rising, a gas-forming reaction between tartaric acid and baking soda.

Most baking powders contain both tartaric acid and sodium hydrogen carbonate, which are kept apart by using starch as a filler. When baking powder is mixed into the moist batter, the acid and sodium hydrogen carbonate dissolve slightly, which allows them to come into contact and react. Carbon dioxide is produced, and the batter rises.

Precipitation Reactions

Formation of an insoluble compound will sometimes occur when a solution containing a particular cation (a positively charged ion) is mixed with another solution containing a particular anion (a negatively charged ion). The solid that separates is called a precipitate.

Compounds having anions such as sulfide (S^{2-}), hydroxide (OH^-), carbonate (CO_3^{2-}), and phosphate (PO_4^{3-}) are often insoluble in water. A precipitate will form if a solution containing one of these anions is added to a solution containing a metal cation such as Fe^{2+}, Cu^{2+}, or Al^{3+}:

$$Fe^{2+}(aq) + 2\ OH^-(aq) \rightarrow Fe(OH)2(s)$$

$$Al^{3+}(aq) + PO_4^{3-}(aq) \rightarrow AlPO^4(s)$$

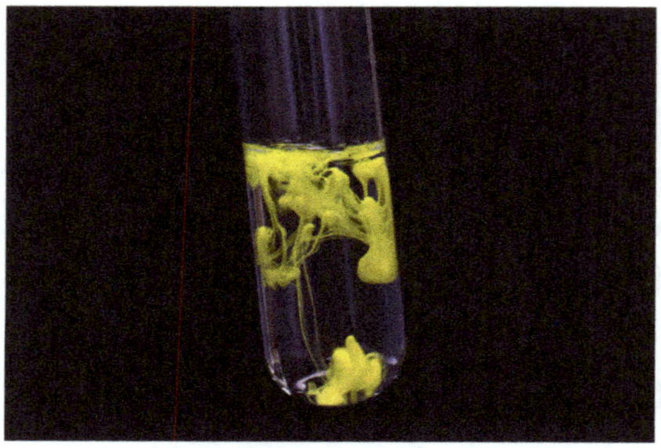

The precipitation of lead(II) iodide (plumbous iodide) results in the formation of solid, yellow matter.

Minerals are water-insoluble compounds. Precipitation reactions in nature can account for mineral formation in many cases, such as with undersea vents called "black smokers" that form metal sulfides.

Classification by Types of Reactants

Two types of reactions involve transfer of a charged species. Oxidation-reduction reactions occur with electron transfer between reagents. In contrast, reactions of acids with bases in water involve proton (H^+) transfer from an acid to a base.

Oxidation-reduction Reactions

Oxidation-reduction (redox) reactions involve the transfer of one or more electrons from a reducing agent to an oxidizing agent. This has the effect of reducing the real or apparent electric charge on an atom in the substance being reduced and of increasing the real or apparent electric charge on an atom in the substance being oxidized. Simple redox reactions include the reactions of an element with oxygen. For example, magnesium burns in oxygen to form magnesium oxide (MgO). The product is an ionic compound, made up of Mg^{2+} and O^{2-} ions. The reaction occurs with each magnesium atom giving up two electrons and being oxidized and each oxygen atom accepting two electrons and being reduced.

Another common redox reaction is one step in the rusting of iron in damp air:

$$2Fe(s) + 2H_2O(l) + O_2(g) \rightarrow 2Fe(OH)_2(s)$$

Here iron metal is oxidized to iron dihydroxide ($Fe(OH)_2$); elemental oxygen (O_2) is the oxidizing agent.

Redox reactions are the source of the energy of batteries. The electric current generated by a battery arises because electrons are transferred from a reducing agent to an oxidizing agent through the external circuitry. In a common dry cell and in alkaline batteries, two electrons per zinc atom are transferred to the oxidizing agent, thereby converting zinc metal to the Zn^{2+} ion. In dry-cell batteries, which are often used in flashlights, the electrons given up by zinc are taken up by ammonium ions (NH_4^+) present in the battery as ammonium chloride (NH_4Cl). In alkaline batteries, which are used in calculators and watches, the electrons are transferred to a metal oxide such as silver oxide (AgO), which is reduced to silver metal in the process.

Acid-base Reactions

Acids and bases are important compounds in the natural world, so their chemistry is central to any discussion of chemical reactions. There are several theories of acid-base behaviour.

The Arrhenius Theory

The Arrhenius theory, named after Swedish physicist Svante August Arrhenius, views an acid as a substance that increases the concentration of the hydronium ion (H_3O^+) in an aqueous solution and a base as a substance that increases the hydroxide ion (OH^-) concentration. Well-known acids include hydrochloric acid (HCl), sulfuric acid (H_2SO_4), nitric acid (HNO_3), and acetic acid (CH_3COOH). Bases includes such common substances as caustic soda (sodium hydroxide, NaOH) and slaked lime (calcium hydroxide, $Ca(OH)_2$). Another common base is ammonia (NH_3), which reacts with water to give a basic solution according to the following balanced equation:

$$NH_3(aq) + H_2O(l) \rightarrow NH_4^+(aq) + OH^-(aq)$$

This reaction occurs to a very small extent; the hydroxide ion concentration is small but measurable.

A large number of natural bases are known, including morphine, cocaine, nicotine, and caffeine; many synthetic drugs are also bases. All of these contain a nitrogen atom bonded to three other groups, and all behave similarly to ammonia in that they can react with water to give a solution containing the hydroxide ion.

Amino acids, a very important class of compounds, are able to function both as acids and as bases. Amino acid molecules contain both acidic (−COOH) and basic ($-NH_2$) sites. In an aqueous solution, amino acids exist in both the molecular form and the so-called "zwitterionic" form, $H_3N + CH_2CO_2^-$. In this structure the nitrogen atom bears a positive charge, and the oxygen atom of the acid group bears a negative charge.

According to the Arrhenius theory, acid-base reactions involve the combination of the hydrogen ion (H^+) and the hydroxide ion to form water. An example is the reaction of aqueous solutions of sodium hydroxide and hydrochloric acid:

$$HCl(aq) + NaOH(aq) \rightarrow NaCl(aq) + H_2O\ (l)$$

The Brønsted-Lowry Theory

A somewhat more general acid-base theory, the Brønsted-Lowry theory, named after Danish chemist Johannes Nicolaus Brønsted and English chemist Thomas Martin Lowry, defines an acid as a proton donor and a base as a proton acceptor. In this theory, the reaction of an acid and base is represented as an equilibrium reaction:

$$\text{acid (1)} + \text{base (2)} \rightleftharpoons \text{base (1)} + \text{acid (2)}$$

(The double arrows, $\rightleftharpoons$, indicate that the products can re-form the reactants in a dynamic process.)

Acid (1) and base (1) are called a conjugate acid-base pair, as are acid (2) and base (2). The advantage of this theory is its predictive capacity. Whether the equilibrium lies toward the reactants (reactant-favoured) or the products (product-favoured) is determined by the relative strengths of the acids and bases.

The Brønsted-Lowry theory is often closely associated with the solvent water. Dissolving an acid in water to form the hydronium ion and the anion of the acid is an acid-base reaction. Acids are classified as strong or weak, depending on whether the equilibrium favours the reactants or products. Hydrochloric acid, a strong acid, ionizes completely in water to form the hydronium and chlorine (Cl^-) ions in a product-favoured reaction.

$$HCl(aq) + H_2O\ (l) \rightarrow H_3O^+(aq) + Cl^-(aq)$$

Using the Brønsted-Lowry theory, the reaction of ammonia and hydrochloric acid in water is represented by the following equation:

$$NH_3(aq) + HCl(aq) \rightarrow NH_4^+(aq) + Cl^-(aq)$$

Hydrochloric acid and the chlorine ion are one conjugate acid-base pair, and the ammonium ion and ammonia are the other. The acid-base reaction is the transfer of the hydrogen ion from the acid (HCl) to the base (NH_3). The equilibrium favours the weaker acid and base, in this case the products. Note that the hydroxide ion does not appear in this equation, a point differentiating the Arrhenius and Brønsted-Lowry theories.

The Lewis Theory

A still broader acid and base theory was proposed by American physical chemist Gilbert Newton Lewis. In the Lewis theory, bases are defined as electron-pair donors and acids as electron-pair acceptors. Acid-base reactions involve the combination of the Lewis acid and base through sharing of the base's electron pair.

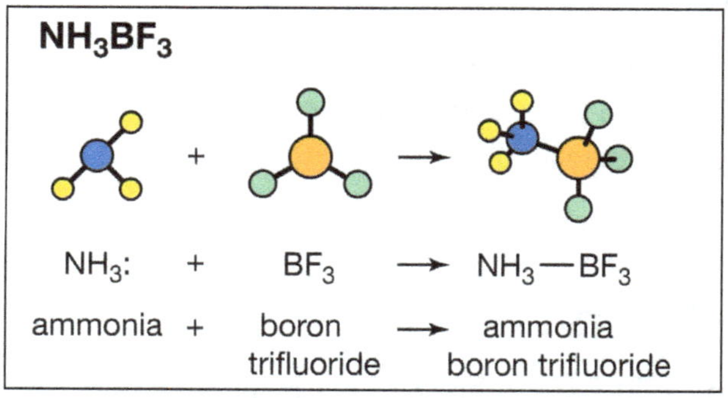

Acid-base reaction with ammonia (NH_3) and boron trifluoride (BF_3) to form ammonia boron trifluoride.

Ammonia is an example of a Lewis base. A pair of electrons located on the nitrogen atom may be used to form a chemical bond to a Lewis acid such as boron trifluoride (BF_3). (In the following equation, the colon represents an electron pair.)

$$H_3N{:} + BF_3 \rightarrow H_3N - BF_3$$

Ammonia, water, and many other Lewis bases react with metal ions to form a group of species known as coordination compounds. The reaction to form these species is another example of a

Lewis acid-base reaction. For example, the light blue colour of a solution of Cu^{2+} ions in water is due to the $[Cu(H_2O)_6]^{2+}$ ion. If ammonia is added to this solution, the water molecules attached to copper are replaced by ammonia molecules, and the beautiful deep blue ion $[Cu(NH_3)_4]^{2+}$ is formed.

Classification by Reaction Outcome

Chemists often classify reactions on the basis of the overall result. Here several commonly encountered reactions are classified. As previously noted, many reactions defy simple classification and may fit in several categories.

Decomposition Reactions

Decomposition reactions are processes in which chemical species break up into simpler parts. Usually, decomposition reactions require energy input. For example, a common method of producing oxygen gas is the decomposition of potassium chlorate ($KClO_3$) by heat:

$$2KClO_3(s) \rightarrow 2KCl(s) + 3O_2(g)$$

Another decomposition reaction is the production of sodium (Na) and chlorine (Cl_2) by electrolysis of molten sodium chloride (NaCl) at high temperature:

$$2NaCl\ (l) \rightarrow 2Na\ (l) + Cl_2(g)$$

A decomposition reaction that was very important in the history of chemistry is the decomposition of mercury oxide (HgO) with heat to give mercury metal (Hg) and oxygen gas. This is the reaction used by 18th-century chemists Carl Wilhelm Scheele, Joseph Priestley, and Antoine-Laurent Lavoisier in their experiments on oxygen:

$$2HgO(s) \rightarrow 2Hg\ (l) + O_2(g)$$

Substitution, Elimination and Addition Reactions

These terms are particularly useful in describing organic reactions. In a substitution reaction, an atom or group of atoms in a molecule is replaced by another atom or group of atoms. For example, methane (CH_4) reacts with chlorine (Cl_2) to produce chloromethane (CH_3Cl), a compound used as a topical anesthetic. In this reaction, a chlorine atom is substituted for a hydrogen atom.

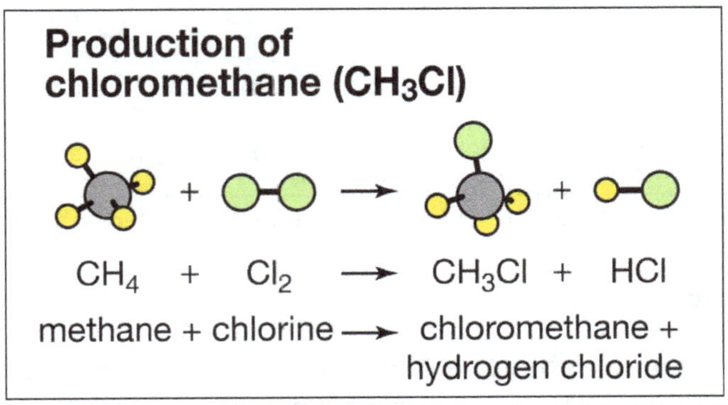

The reaction of methane (CH_4) and chlorine (Cl_2) to form chloromethane (CH_3Cl) and hydrochloric acid (HCl).

Substitution reactions are widely used in industrial chemistry. For example, substituting two of the chlorine atoms on chloroform ($CHCl_3$) with fluorine atoms produces chlorodifluoromethane ($CHClF_2$). This product undergoes a further reaction when heated strongly:

$$2CHClF_2(g) \rightarrow F_2C{=}CF_2(g) + 2HCl(g)$$

This latter reaction is an example of an elimination reaction, a hydrogen atom and a chlorine atom being eliminated from the starting material as hydrochloric acid (HCl). The other product is tetrafluoroethylene, a precursor to the polymer known commercially as Teflon.

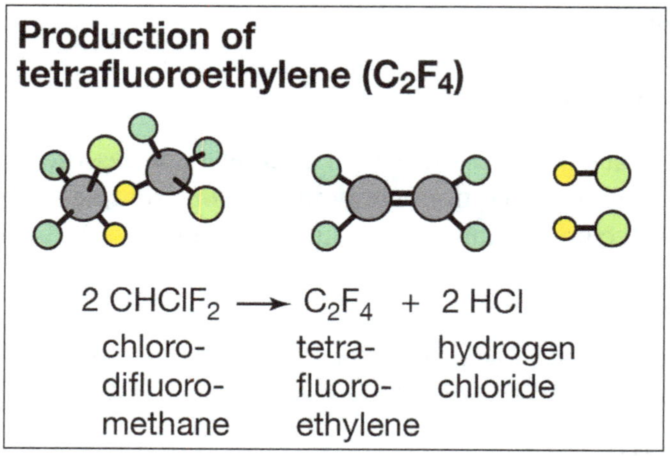

Elimination reaction in which two molecules of chlorodifluoromethane ($CHClF_2$) produce tetrafluoroethylene (C_2F_4) and two molecules of hydrochloric acid (HCl).

Addition reactions are the opposite of elimination reactions. As the name implies, one molecule is added to another. An example is the common industrial preparation of ethanol (CH_3CH_2OH). Historically, this compound was made by fermentation. However, since the early 1970s, it has also been made commercially by the addition of water to ethylene:

$$C_2H_4 + H_2O \rightarrow CH_3CH_2OH$$

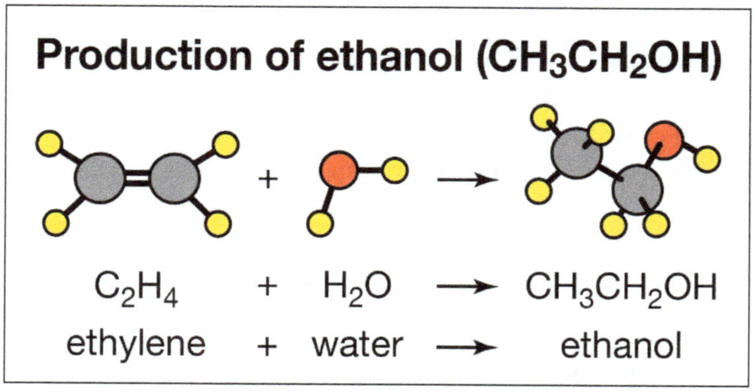

Addition reaction of ethylene (C_2H_4) with water (H_2O) to produce ethanol (C_2H_5OH).

Polymerization Reactions

Polymers are high-molecular-weight compounds, fashioned by the aggregation of many smaller molecules called monomers. The plastics that have so changed society and the natural and synthetic fibres used in clothing are polymers. There are two basic ways to form polymers: (a) linking small molecules together, a type of addition reaction, and (b) combining two molecules (of the same or different type) with the elimination of a stable small molecule such as water. This latter type of polymerization combines addition and elimination reactions and is called a condensation reaction .

An example of the first type of reaction is the union of thousands of ethylene molecules that gives polyethylene,

$$nH_2C{=}CH_2 \rightarrow [-CH_2CH_2-]_n$$

Other addition polymers include polypropylene (made by polymerizing $H_2C{=}CHCH_3$), polystyrene (from $H_2C{=}CH\ C_6H_5$), and polyvinyl chloride (from $H_2C{=}CHCl$).

Starch and cellulose are examples of the second type of polymer. These are members of a class of compounds called carbohydrates, substances with formulas that are multiples of the simple formula CH_2O. Both starch and cellulose are polymers of glucose, a sugar with the formula $C_6H_{12}O_6$. In both starch and cellulose, molecules of glucose are joined together with concomitant elimination of a molecule of water for every linkage formed:

$$nC_6H_{12}O_6 \rightarrow -[-C_6H_{10}O_5-]-n + nH_2O$$

The synthetic material nylon is another example of this type of polymer. Water and a polymer (nylon-6,6) are formed when an organic acid and an amine (a compound derived from ammonia) combine.

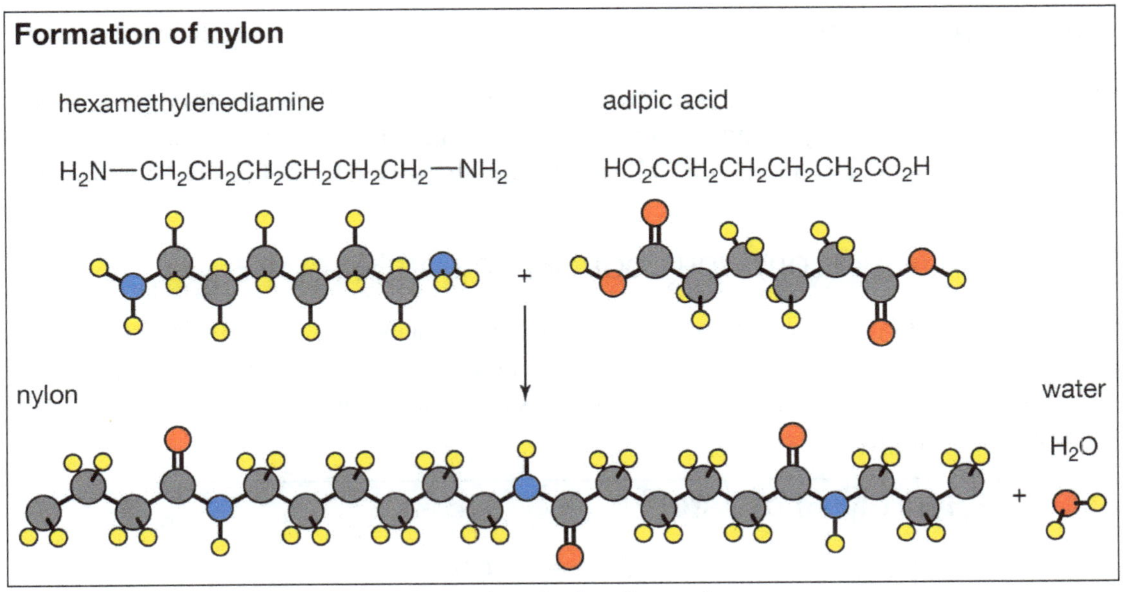

Formation of nylon (bottom).

The natural fibres of proteins such as hair, wool, and silk are also polymers that contain the repeating unit (-CHRCONH-), where R is a group of atoms attached to the main polymer. These form by joining amino acids with the elimination of a water molecule for each CONH or peptide linkage formed; for example, the structure of the tripeptide chain is formed from three units of the amino acid glycine ($NH_2CH_2CO_2H$).

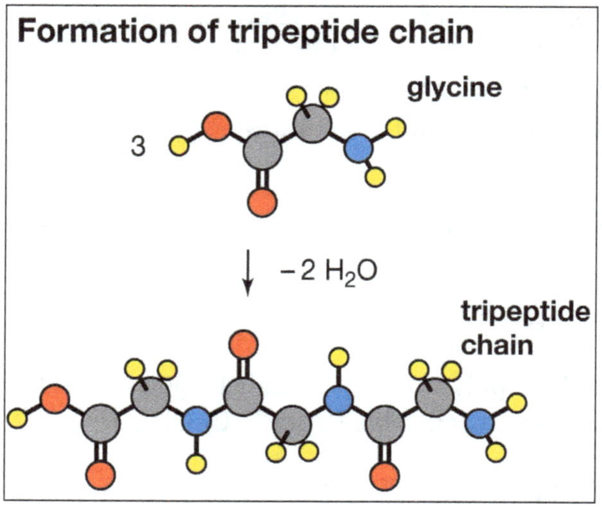

Condensation reaction in which three molecules of the amino acid glycine produce a tripeptide chain, with the elimination of two molecules of water (H_2O).

Solvolysis and Hydrolysis

A solvolysis reaction is one in which the solvent is also a reactant. Solvolysis reactions are generally named after the specific solvent—for example, the term hydrolysis when water is involved. If a compound is represented by the formula AB (in which A and B are atoms or groups of atoms) and water is represented by the formula HOH, the hydrolysis reaction may be represented by the reversible chemical reaction:

$$AB + HOH \rightleftharpoons AH + BOH.$$

Hydrolysis of an organic compound is illustrated by the reaction of water with esters. Esters have the general formula RCOOR′, R and R′ being combining groups (such as CH_3). The hydrolysis of an ester produces an acid and an alcohol. The equation for the reaction of methyl acetate and water is:

$$CH_3COOCH_3(aq) + H_2O(l) \rightarrow CH_3COOH(aq) + CH_3OH(aq).$$

Hydrolysis reactions play an important role in chemical processes that occur in living organisms. Proteins are hydrolyzed to amino acids, fats to fatty acids and glycerol, and starches and complex sugars to simple sugars. In most instances, the rates of these processes are enhanced by the presence of enzymes, biological catalysts.

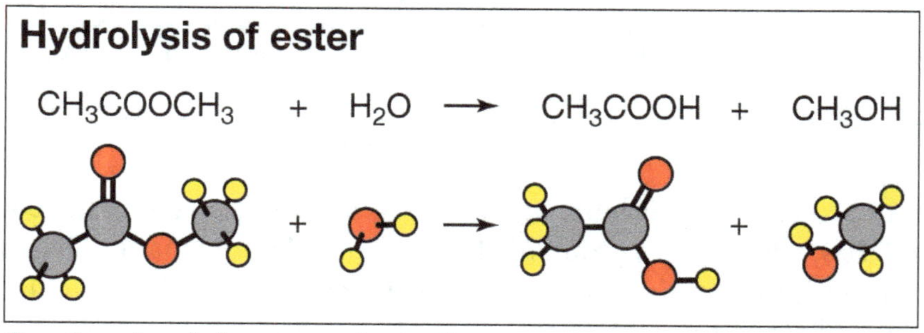

Reaction of methyl acetate and water demonstrating the hydrolysis of an ester.

Hydrolysis reactions are also important to acid-base behaviour. Anions of weak acids dissolve in water to give basic solutions, as in the hydrolysis of the acetate ion, CH_3COO^-.

$$CH_3COO^-(aq) + H_2O\,(l) \rightarrow CH_3COOH(aq) + OH^-(aq)$$

Although this is a reactant-favoured reaction, it occurs to an extent sufficient to cause a solution containing the acetate ion to exhibit basic properties (e.g., turning red litmus paper blue).

Hydrolysis reactions account for the basic character of many common substances. Salts of the borate, phosphate, and carbonate ions, for example, give basic solutions that have long been used for cleaning purposes. Many food products also contain basic anions such as tartrate and citrate ions.

Classification by Reaction Mechanism

Reaction mechanisms provide details on how atoms are shuffled and reassembled in the formation of products from reactants. Chain and photolysis reactions are named on the basis of the mechanism of the process.

Chemical Reaction Model

Chemical reaction models transform physical knowledge into a mathematical formulation that can be utilized in computational simulation of practical problems in chemical engineering. Computer simulation provides the flexibility to study chemical processes under a wide range of conditions. Modeling of a chemical reaction involves solving conservation equations describing convection, diffusion, and reaction source for each component species.

Species Transport Equation

$$\frac{\partial(\rho Y_i)}{\partial t} + \nabla\cdot(\rho\vec{v}Y_i) = -\nabla\cdot\vec{J}_i + R_i + S_i$$

R_i is the net rate of production of species i by chemical reaction and S_i is the rate of creation by addition from the dispersed phase and the user defined source. J_i is the diffusion flux of species i, which arises due to concentration gradients and differs in both laminar and turbulent flows. In turbulent flows, computational fluid dynamics also considers the effects of turbulent diffusivity. The net source of chemical species i due to reaction, R_i which appeared as the source term in the species transport equation is computed as the sum of the reaction sources over the N_R reactions among the species.

Reaction Models

These reaction rates R can be calculated by following models:

1. Laminar finite rate model
2. Eddy dissipation model
3. Eddy dissipation concept

Laminar Finite Rate Model

The laminar finite rate model computes the chemical source terms using the Arrhenius expressions and ignores turbulence fluctuations. This model provides with the exact solution for laminar flames but gives inaccurate solution for turbulent flames, in which turbulence highly affects the chemistry reaction rates, due to highly non-linear Arrhenius chemical kinetics. However this model may be accurate for combustion with small turbulence fluctuations, for example supersonic flames.

Eddy Dissipation Model

The eddy dissipation model, based on the work of Magnussen and Hjertager, is a turbulent-chemistry reaction model. Most fuels are fast burning and the overall rate of reaction is controlled by turbulence mixing. In the non-premixed flames, turbulence slowly mixes the fuel and oxidizer into the reaction zones where they burn quickly. In premixed flames the turbulence slowly mixes cold reactants and hot products into the reaction zones where reaction occurs rapidly. In such cases the combustion is said to be mixing-limited, and the complex and often unknown chemical kinetics can be safely neglected. In this model, the chemical reaction is governed by large eddy mixing time scale. Combustion initiates whenever there is turbulence present in the flow. It does not need an ignition source to initiate the combustion. This type of model is valid for the non-premixed combustion, but for the premixed flames the reactant is assumed to burn at the moment it enters the computation model, which is a shortcoming of this model as in practice the reactant needs some time to get to the ignition temperature to initiate the combustion.

Eddy Dissipation Concept

The eddy dissipation concept (EDC) model is an extension of the eddy dissipation model to include detailed chemical mechanism in turbulent flows. The EDC model attempts to incorporate the significance of fine structures in a turbulent reacting flow in which combustion is important. EDC has been proven efficient without the need for changing the constants for a great variety of premixed and diffusion controlled combustion problems, both where the chemical kinetics is faster than the overall fine structure mixing as well as in cases where the chemical kinetics has a dominating influence.

Chain Reaction

A chain reaction is a sequence of reactions where a reactive product or by-product causes additional reactions to take place. In a chain reaction, positive feedback leads to a self-amplifying chain of events.

Chain reactions are one way that systems which are not in thermodynamic equilibrium can release energy or increase entropy in order to reach a state of higher entropy. For example, a system may not be able to reach a lower energy state by releasing energy into the environment, because it is hindered or prevented in some way from taking the path that will result in the energy release. If a reaction results in a small energy release making way for more energy releases in an expanding chain, then the system will typically collapse explosively until much or all of the stored energy has been released.

A macroscopic metaphor for chain reactions is thus a snowball causing a larger snowball until finally an avalanche results ("snowball effect"). This is a result of stored gravitational potential energy seeking a path of release over friction. Chemically, the equivalent to a snow avalanche is a spark causing a forest fire. In nuclear physics, a single stray neutron can result in a prompt critical event, which may finally be energetic enough for a nuclear reactor meltdown or (in a bomb) a nuclear explosion. Numerous chain reactions can be represented by a mathematical model based on Markov chains.

Chemical Chain Reactions

In 1913, the German chemist Max Bodenstein first put forth the idea of chemical chain reactions. If two molecules react, not only molecules of the final reaction products are formed, but also some unstable molecules which can further react with the parent molecules with a far larger probability than the initial reactants (In the new reaction, further unstable molecules are formed besides the stable products, and so on).

In 1918, Walther Nernst proposed that the photochemical reaction between hydrogen and chlorine is a chain reaction in order to explain what's known as the *quantum yield* phenomena. This means that one photon of light is responsible for the formation of as many as 10^6 molecules of the product HCl. Nernst suggested that the photon dissociates a Cl_2 molecule into two Cl atoms which each initiate a long chain of reaction steps forming HCl.

In 1923, Danish and Dutch scientists Christian Christiansen and Hendrik Anthony Kramers, in an analysis of the formation of polymers, pointed out that such a chain reaction need not start with a molecule excited by light, but could also start with two molecules colliding violently due to thermal energy as previously proposed for initiation of chemical reactions by van' t Hoff.

Christiansen and Kramers also noted that if, in one link of the reaction chain, two or more unstable molecules are produced, the reaction chain would branch and grow. The result is in fact an exponential growth, thus giving rise to explosive increases in reaction rates, and indeed to chemical explosions themselves. This was the first proposal for the mechanism of chemical explosions.

A quantitative chain chemical reaction theory was created later on by Soviet physicist Nikolay Semyonov in 1934. Semyonov shared the Nobel Prize in 1956 with Sir Cyril Norman Hinshelwood, who independently developed many of the same quantitative concepts.

Typical Steps

The main types of steps in chain reaction are of the following types.

- Initiation (formation of active particles or chain carriers, often free radicals, in either a thermal or a photochemical step).
- Propagation (may comprise several elementary steps in a cycle, where the active particle through reaction forms another active particle which continues the reaction chain by entering the next elementary step). In effect the active particle serves as a catalyst for the overall reaction of the propagation cycle. Particular cases are:
    - Chain branching (a propagation step which forms more new active particles than enter the step);

- Chain transfer (a propagation step in which the active particle is a growing polymer chain which reacts to form an inactive polymer whose growth is terminated and an active small particle (such as a radical), which may then react to form a new polymer chain).

- Termination (elementary step in which the active particle loses its activity; e. g. by recombination of two free radicals).

The *chain length* is defined as the average number of times the propagation cycle is repeated, and equals the overall reaction rate divided by the initiation rate.

Some chain reactions have complex rate equations with fractional order or mixed order kinetics.

Example: The Hydrogen-bromine Reaction

The reaction $H_2 + Br_2 \rightarrow 2\ HBr$ proceeds by the following mechanism:

- Initiation:

  $Br_2 \rightarrow 2\ Br\bullet$ (thermal) or $Br_2 + h\nu \rightarrow 2\ Br\bullet$ (photochemical)

  each Br atom is a free radical, indicated by the symbol « • » representing an unpaired electron.

- Propagation (here a cycle of two steps):

  $Br\bullet + H_2 \rightarrow HBr + H\bullet$

  $H\bullet + Br_2 \rightarrow HBr + Br\bullet$

  the sum of these two steps corresponds to the overall reaction $H_2 + Br_2 \rightarrow 2\ HBr$, with catalysis by Br• which participates in the first step and is regenerated in the second step.

- Retardation (inhibition):

  $H\bullet + HBr \rightarrow H_2 + Br\bullet$

  this step is specific to this example, and corresponds to the first propagation step in reverse.

- Termination $2\ Br\bullet \rightarrow Br_2$

  recombination of two radicals, corresponding in this example to initiation in reverse.

As can be explained using the steady-state approximation, the thermal reaction has an initial rate of fractional order (3/2), and a complete rate equation with a two-term denominator (mixed-order kinetics).

Further Chemical Examples

- The reaction $2\ H_2 + O_2 \rightarrow 2\ H_2O$ provides an example of chain branching. The propagation is a sequence of two steps whose net effect is to replace an H atom by another H atom plus two OH radicals. This leads to an explosion under certain conditions of temperature and pressure.

- $H + O_2 \rightarrow OH + O$
- $O + H_2 \rightarrow OH + H$

- In chain-growth polymerization, the propagation step corresponds to the elongation of the growing polymer chain. Chain transfer corresponds to transfer of the activity from this growing chain, whose growth is terminated, to another molecule which may be a second growing polymer chain. For polymerization, the kinetic chain length defined above may differ from the degree of polymerization of the product macromolecule.
- Polymerase chain reaction, a technique used in molecular biology to amplify (make many copies of) a piece of DNA by *in vitro* enzymatic replication using a DNA polymerase.

Electron Avalanche in Gases

An electron avalanche happens between two unconnected electrodes in a gas when an electric field exceeds a certain threshold. Random thermal collisions of gas atoms may result in a few free electrons and positively charged gas ions, in a process called impact ionization. Acceleration of these free electrons in a strong electric field causes them to gain energy, and when they impact other atoms, the energy causes release of new free electrons and ions (ionization), which fuels the same process. If this process happens faster than it is naturally quenched by ions recombining, the new ions multiply in successive cycles until the gas breaks down into a plasma and current flows freely in a discharge.

Electron avalanches are essential to the dielectric breakdown process within gases. The process can culminate in corona discharges, streamers, leaders, or in a spark or continuous electric arc that completely bridges the gap. The process may extend huge sparks — streamers in lightning discharges propagate by formation of electron avalanches created in the high potential gradient ahead of the streamers' advancing tips. Once begun, avalanches are often intensified by the creation of photoelectrons as a result of ultraviolet radiation emitted by the excited medium's atoms in the aft-tip region. The extremely high temperature of the resulting plasma cracks the surrounding gas molecules and the free ions recombine to create new chemical compounds.

The process can also be used to detect radiation that initiates the process, as the passage of a single particles can be amplified to large discharges. This is the mechanism of a Geiger counter and also the visualization possible with a spark chamber and other wire chambers.

Avalanche Breakdown in Semiconductors

An avalanche breakdown process can happen in semiconductors, which in some ways conduct electricity analogously to a mildly ionized gas. Semiconductors rely on free electrons knocked out of the crystal by thermal vibration for conduction. Thus, unlike metals, semiconductors become better conductors the higher the temperature. This sets up conditions for the same type of positive feedback—heat from current flow causes temperature to rise, which increases charge carriers, lowering resistance, and causing more current to flow. This can continue to the point of complete breakdown of normal resistance at a semiconductor junction, and failure of the device (this may be temporary or permanent depending on whether there is physical damage to the crystal). Certain devices, such as avalanche diodes, deliberately make use of the effect.

Nuclear Chain Reaction

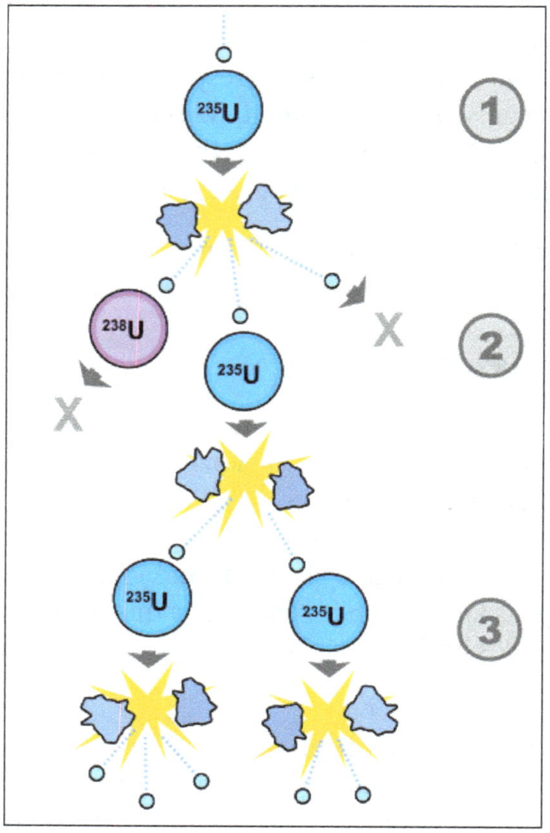

A possible nuclear fission chain reaction. 1. A uranium-235 atom absorbs a neutron, and fissions into two new atoms (fission fragments), releasing three new neutrons and a large amount of binding energy. 2. One of those neutrons is absorbed by an atom of uranium-238, and does not continue the reaction. Another neutron leaves the system without being absorbed. However, one neutron does collide with an atom of uranium-235, which then fissions and releases two neutrons and more binding energy. 3. Both of those neutrons collide with uranium-235 atoms, each of which fissions and releases a few neutrons, which can then continue the reaction.

A nuclear chain reaction occurs when one single nuclear reaction causes an average of one or more subsequent nuclear reactions, this leading to the possibility of a self-propagating series of these reactions. The specific nuclear reaction may be the fission of heavy isotopes (e.g., uranium-235, ^{235}U). The nuclear chain reaction releases several million times more energy per reaction than any chemical reaction.

Fission Chain Reaction

Fission chain reactions occur because of interactions between neutrons and fissile isotopes (such as ^{235}U). The chain reaction requires both the release of neutrons from fissile isotopes undergoing nuclear fission and the subsequent absorption of some of these neutrons in fissile isotopes. When an atom undergoes nuclear fission, a few neutrons (the exact number depends on uncontrollable and unmeasurable factors; the expected number depends on several factors, usually between 2.5 and 3.0) are ejected from the reaction. These free neutrons will then interact with the surrounding medium, and if more fissile fuel is present, some may be absorbed and cause more fissions. Thus, the cycle repeats to give a reaction that is self-sustaining.

Nuclear power plants operate by precisely controlling the rate at which nuclear reactions occur, on the other hand, nuclear weapons are specifically engineered to produce a reaction that is so fast and intense it cannot be controlled after it has started. When properly designed, this uncontrolled reaction will lead to an explosive energy release.

Nuclear Fission Fuel

Nuclear weapons employ high quality, highly enriched fuel exceeding the critical size and geometry (critical mass) necessary in order to obtain an explosive chain reaction. The fuel for energy purposes, such as in a nuclear fission reactor, is very different, usually consisting of a low-enriched oxide material (e.g. UO_2).

Fission Reaction Products

When a fissile atom undergoes nuclear fission, it breaks into two or more fission fragments. Also, several free neutrons, gamma rays, and neutrinos are emitted, and a large amount of energy is released. The sum of the rest masses of the fission fragments and ejected neutrons is less than the sum of the rest masses of the original atom and incident neutron (of course the fission fragments are not at rest). The mass difference is accounted for in the release of energy according to the equation $E=\Delta mc^2$:

$$\text{mass of released energy} = \frac{E}{c^2} = m_{\text{original}} - m_{\text{final}}$$

Due to the extremely large value of the speed of light, c, a small decrease in mass is associated with a tremendous release of active energy (for example, the kinetic energy of the fission fragments). This energy (in the form of radiation and heat) carries the missing mass, when it leaves the reaction system (total mass, like total energy, is always conserved). While typical chemical reactions release energies on the order of a few eVs (e.g. the binding energy of the electron to hydrogen is 13.6 eV), nuclear fission reactions typically release energies on the order of hundreds of millions of eVs.

Two typical fission reactions are shown below with average values of energy released and number of neutrons ejected:

$$^{235}\text{U} + \text{neutronn} \rightarrow \quad \text{fission fragments} + 2.4 \text{ neutrons} + 192.9 \text{ MeV}$$

$$^{239}\text{Pu} + \text{neutronn} \rightarrow \quad \text{fission fragments} + 2.9 \text{ neutrons} + 198.5 \text{ MeV}$$

Note that these equations are for fissions caused by slow-moving (thermal) neutrons. The average energy released and number of neutrons ejected is a function of the incident neutron speed. Also, note that these equations exclude energy from neutrinos since these subatomic particles are extremely non-reactive and, therefore, rarely deposit their energy in the system.

Timescales of Nuclear Chain Reactions

Prompt Neutron Lifetime

The prompt neutron lifetime, *l*, is the average time between the emission of neutrons and either their absorption in the system or their escape from the system. The neutrons that occur directly from fission are called "prompt neutrons," and the ones that are a result of radioactive decay of fission fragments are called "delayed neutrons". The term lifetime is used because the emission of a neutron is often considered its "birth," and the subsequent absorption is considered its "death". For thermal (slow-neutron) fission reactors, the typical prompt neutron lifetime is on the order of 10^{-4} seconds, and for fast fission reactors, the prompt neutron lifetime is on the order of 10^{-7} seconds.These extremely short lifetimes mean that in 1 second, 10,000 to 10,000,000 neutron lifetimes can pass. The *average* (also referred to as the *adjoint unweighted*) prompt neutron lifetime takes into account all prompt neutrons regardless of their importance in the reactor core; the *effective* prompt neutron lifetime (referred to as the *adjoint weighted* over space, energy, and angle) refers to a neutron with average importance.

Mean Generation Time

The mean generation time, Λ, is the average time from a neutron emission to a capture that results in fission. The mean generation time is different from the prompt neutron lifetime because the mean generation time only includes neutron absorptions that lead to fission reactions (not other absorption reactions). The two times are related by the following formula:

$$\Lambda = \frac{l}{k}$$

In this formula, k is the effective neutron multiplication factor.

Effective Neutron Multiplication Factor

The six factor formula effective neutron multiplication factor, *k*, is the average number of neutrons from one fission that cause another fission. The remaining neutrons either are absorbed in non-fission reactions or leave the system without being absorbed. The value of *k* determines how a nuclear chain reaction proceeds:

- $k < 1$ (subcriticality): The system cannot sustain a chain reaction, and any beginning of a chain reaction dies out over time. For every fission that is induced in the system, an average *total* of $1/(1-k)$ fissions occur.

- $k = 1$ (criticality): Every fission causes an average of one more fission, leading to a fission (and power) level that is constant. Nuclear power plants operate with $k = 1$ unless the power level is being increased or decreased.

- $k > 1$ (supercriticality): For every fission in the material, it is likely that there will be "*k*" fissions after the next mean generation time (Λ). The result is that the number of fission reactions increases exponentially, according to the equation $e^{(k-1)t/\Lambda}$, where t is the elapsed time. Nuclear weapons are designed to operate under this state. There are two subdivisions of supercriticality: prompt and delayed.

When describing kinetics and dynamics of nuclear reactors, and also in the practice of reactor operation, the concept of reactivity is used, which characterizes the deflection of reactor from the critical state. $\rho=(k-1)/k$. InHour is a unit of reactivity of a nuclear reactor.

In a nuclear reactor, *k* will actually oscillate from slightly less than 1 to slightly more than 1, due primarily to thermal effects (as more power is produced, the fuel rods warm and thus expand, lowering their capture ratio, and thus driving *k* lower). This leaves the average value of *k* at exactly 1. Delayed neutrons play an important role in the timing of these oscillations.

In an infinite medium, the multiplication factor may be described by the four factor formula; in a non-infinite medium, the multiplication factor may be described by the six factor formula.

Prompt and Delayed Supercriticality

Not all neutrons are emitted as a direct product of fission; some are instead due to the radioactive decay of some of the fission fragments. The neutrons that occur directly from fission are called "prompt neutrons," and the ones that are a result of radioactive decay of fission fragments are called "delayed neutrons". The fraction of neutrons that are delayed is called β, and this fraction is typically less than 1% of all the neutrons in the chain reaction.

The delayed neutrons allow a nuclear reactor to respond several orders of magnitude more slowly than just prompt neutrons would alone. Without delayed neutrons, changes in reaction rates in nuclear reactors would occur at speeds that are too fast for humans to control.

The region of supercriticality between $k = 1$ and $k = 1/(1-\beta)$ is known as delayed supercriticality (or delayed criticality). It is in this region that all nuclear power reactors operate. The region of supercriticality for $k > 1/(1-\beta)$ is known as prompt supercriticality (or prompt criticality), which is the region in which nuclear weapons operate.

The change in k needed to go from critical to prompt critical is defined as a dollar.

Nuclear Weapons Application of Neutron Multiplication

Nuclear fission weapons require a mass of fissile fuel that is prompt supercritical.

For a given mass of fissile material the value of k can be increased by increasing the density. Since the probability per distance traveled for a neutron to collide with a nucleus is proportional to the material density, increasing the density of a fissile material can increase k. This concept is utilized in the implosion method for nuclear weapons. In these devices, the nuclear chain reaction begins after increasing the density of the fissile material with a conventional explosive.

In the gun-type fission weapon two subcritical pieces of fuel are rapidly brought together. The value of k for a combination of two masses is always greater than that of its components. The magnitude of the difference depends on distance, as well as the physical orientation. The value of k can also be increased by using a neutron reflector surrounding the fissile material.

Once the mass of fuel is prompt supercritical, the power increases exponentially. However, the exponential power increase cannot continue for long since k decreases when the amount of fission material that is left decreases (i.e. it is consumed by fissions). Also, the geometry and density are expected to change during detonation since the remaining fission material is torn apart from the explosion.

Predetonation

Detonation of a nuclear weapon involves bringing fissile material into its optimal supercritical state very rapidly. During part of this process, the assembly is supercritical, but not yet in an optimal state for a chain reaction. Free neutrons, in particular from spontaneous fissions, can cause the device to undergo a preliminary chain reaction that destroys the fissile material before it is ready to produce a large explosion, which is known as predetonation.

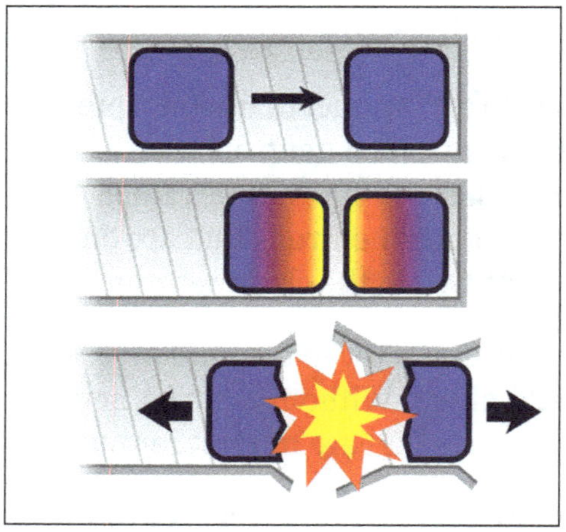

If two pieces of subcritical material are not brought together fast enough, nuclear predetonation can occur, whereby a smaller explosion than expected will blow the bulk of the material apart.

To keep the probability of predetonation low, the duration of the non-optimal assembly period is minimized and fissile and other materials are used which have low spontaneous fission rates. In fact, the combination of materials has to be such that it is unlikely that there is even a single spontaneous fission during the period of supercritical assembly. In particular, the gun method cannot be used with plutonium.

Nuclear Power Plants and Control of Chain Reactions

Chain reactions naturally give rise to reaction rates that grow (or shrink) exponentially, whereas a nuclear power reactor needs to be able to hold the reaction rate reasonably constant. To maintain this control, the chain reaction criticality must have a slow enough time-scale to permit intervention by additional effects (e.g., mechanical control rods or thermal expansion). Consequently, all nuclear power reactors (even fast-neutron reactors) rely on delayed neutrons for their criticality. An operating nuclear power reactor fluctuates between being slightly subcritical and slightly delayed-supercritical, but must always remain below prompt-critical.

It is impossible for a nuclear power plant to undergo a nuclear chain reaction that results in an explosion of power comparable with a nuclear weapon, but even low-powered explosions due to uncontrolled chain reactions, that would be considered “fizzles” in a bomb, may still cause considerable damage and meltdown in a reactor. For example, the Chernobyl disaster involved a runaway chain reaction but the result was a low-powered steam explosion from the relatively small release of heat, as compared with a bomb. However, the reactor complex was destroyed by the heat, as well as by ordinary burning of the graphite exposed to air. Such steam explosions would be typical of the very diffuse assembly of materials in a nuclear reactor, even under the worst conditions.

In addition, other steps can be taken for safety. For example, power plants licensed in the United States require a negative void coefficient of reactivity (this means that if water is removed from the reactor core, the nuclear reaction will tend to shut down, not increase). This eliminates the possibility of the type of accident that occurred at Chernobyl (which was due to a positive void coefficient). However, nuclear reactors are still capable of causing smaller explosions even after complete shutdown, such as was the case of the Fukushima Daiichi nuclear disaster. In such cases, residual decay heat from the core may cause high temperatures if there is loss of coolant flow, even a day after the chain reaction has been shut down. This may cause a chemical reaction between water and fuel that produces hydrogen gas which can explode after mixing with air, with severe contamination consequences, since fuel rod material may still be exposed to the atmosphere from this process. However, such explosions do not happen during a chain reaction, but rather as a result of energy from radioactive beta decay, after the fission chain reaction has been stopped.

Photodissociation

Photodissociation, photolysis, or photodecomposition is a chemical reaction in which a chemical compound is broken down by photons. It is defined as the interaction of one or more photons with one target molecule. Photodissociation is not limited to visible light. Any photon with sufficient energy can affect the chemical bonds of a chemical compound. Since a photon's energy is inversely proportional to its wavelength, electromagnetic waves with the energy of visible light or higher, such as ultraviolet light, x-rays and gamma rays are usually involved in such reactions.

Photolysis in Photosynthesis

Photolysis is part of the light-dependent reactions of photosynthesis. The general reaction of photosynthetic photolysis can be given as:

$$H_2A + 2\text{ photons (light)} \rightarrow 2\,e^- + 2\,H^+ + A$$

The chemical nature of "A" depends on the type of organism. In purple sulfur bacteria, hydrogen sulfide (H_2S) is oxidized to sulfur (S). In oxygenic photosynthesis, water (H_2O) serves as a substrate for photolysis resulting in the generation of diatomic oxygen (O_2). This is the process which returns oxygen to Earth's atmosphere. Photolysis of water occurs in the thylakoids of cyanobacteria and the chloroplasts of green algae and plants.

Energy Transfer Models

The conventional, semi-classical, model describes the photosynthetic energy transfer process as one in which excitation energy hops from light-capturing pigment molecules to reaction center molecules step-by-step down the molecular energy ladder.

The effectiveness of photons of different wavelengths depends on the absorption spectra of the photosynthetic pigments in the organism. Chlorophylls absorb light in the violet-blue and red parts of the spectrum, while accessory pigments capture other wavelengths as well. The phycobilins of red algae absorb blue-green light which penetrates deeper into water than red light, enabling them to photosynthesize in deep waters. Each absorbed photon causes the formation of an exciton (an electron excited to a higher energy state) in the pigment molecule. The energy of the exciton is transferred to a chlorophyll molecule (P680, where P stands for pigment and 680 for its absorption maximum at 680 nm) in the reaction center of photosystem II via resonance energy transfer. P680 can also directly absorb a photon at a suitable wavelength.

Photolysis during photosynthesis occurs in a series of light-driven oxidation events. The energized electron (exciton) of P680 is captured by a primary electron acceptor of the photosynthetic electron transfer chain and thus exits photosystem II. In order to repeat the reaction, the electron in the reaction center needs to be replenished. This occurs by oxidation of water in the case of oxygenic photosynthesis. The electron-deficient reaction center of photosystem II (P680*) is the strongest biological oxidizing agent yet discovered, which allows it to break apart molecules as stable as water.

The water-splitting reaction is catalyzed by the oxygen evolving complex of photosystem II. This protein-bound inorganic complex contains four manganese ions, plus calcium and chloride ions as cofactors. Two water molecules are complexed by the manganese cluster, which then undergoes a series of four electron removals (oxidations) to replenish the reaction center of photosystem II. At the end of this cycle, free oxygen (O_2) is generated and the hydrogen of the water molecules has been converted to four protons released into the thylakoid lumen (Dolai's S-state diagrams).

These protons, as well as additional protons pumped across the thylakoid membrane coupled with the electron transfer chain, form a proton gradient across the membrane that drives photophosphorylation and thus the generation of chemical energy in the form of adenosine triphosphate (ATP). The electrons reach the P700 reaction center of photosystem I where they are energized again by light. They are passed down another electron transfer chain and finally combine with the coenzyme $NADP^+$ and protons outside the thylakoids to form NADPH. Thus, the net oxidation reaction of water photolysis can be written as:

$$2\ H_2O + 2\ NADP^+ + 8\ \text{photons (light)} \rightarrow 2\ NADPH + 2\ H^+ + O_2$$

The free energy change (ΔG) for this reaction is 102 kilocalories per mole. Since the energy of light at 700 nm is about 40 kilocalories per mole of photons, approximately 320 kilocalories of light energy are available for the reaction. Therefore, approximately one-third of the available light energy is captured as NADPH during photolysis and electron transfer. An equal amount of ATP is generated by the resulting proton gradient. Oxygen as a byproduct is of no further use to the reaction and thus released into the atmosphere.

Quantum Models

In 2007 a quantum model was proposed by Graham Fleming and his co-workers which includes the possibility that photosynthetic energy transfer might involve quantum oscillations, explaining its unusually high efficiency.

According to Fleming there is direct evidence that remarkably long-lived wavelike electronic quantum coherence plays an important part in energy transfer processes during photosynthesis, which can explain the extreme efficiency of the energy transfer because it enables the system to sample all the potential energy pathways, with low loss, and choose the most efficient one. This claim has, however, since been proven wrong in several publications.

Photoinduced Proton Transfer

Photoacids are molecules that upon light absorption undergo a proton transfer to form the photobase:

$$AH \xrightarrow{h\nu} A^- + H^+$$

In these reactions the dissociation occurs in the electronically excited state. After proton transfer and relaxation to the electronic ground state, the proton and acid recombine to form the photoacid again.

Photoacids are a convenient source to induce pH jumps in ultrafast laser spectroscopy experiments.

Photolysis in the Atmosphere

Photolysis occurs in the atmosphere as part of a series of reactions by which primary pollutants such as hydrocarbons and nitrogen oxides react to form secondary pollutants such as peroxyacyl nitrates.

The two most important photodissociaton reactions in the troposphere are firstly:

$$O_3 + h\nu \rightarrow O_2 + O(^1D) \quad \lambda < 320 \text{ nm}$$

which generates an excited oxygen atom which can react with water to give the hydroxyl radical:

$$O(^1D) + H_2O \rightarrow 2\ ^{\bullet}OH$$

The hydroxyl radical is central to atmospheric chemistry as it initiates the oxidation of hydrocarbons in the atmosphere and so acts as a detergent.

Secondly the reaction:

$$NO_2 + h\nu \rightarrow NO + O$$

is a key reaction in the formation of tropospheric ozone.

The formation of the ozone layer is also caused by photodissociation. Ozone in the Earth's stratosphere is created by ultraviolet light striking oxygen molecules containing two oxygen atoms (O_2), splitting them into individual oxygen atoms (atomic oxygen). The atomic oxygen then combines with unbroken O_2 to create ozone, O_3. In addition, photolysis is the process by which CFCs are broken down in the upper atmosphere to form ozone-destroying chlorine free radicals.

Astrophysics

In astrophysics, photodissociation is one of the major processes through which molecules are broken down (but new molecules are being formed). Because of the vacuum of the interstellar medium, molecules and free radicals can exist for a long time. Photodissociation is the main path by which molecules are broken down. Photodissociation rates are important in the study of the composition of interstellar clouds in which stars are formed.

Examples of photodissociation in the interstellar medium are (*hv* is the energy of a single photon of frequency *v*):

$$H_2O \xrightarrow{hv} H + OH$$

$$CH_4 \xrightarrow{hv} CH_3 + H$$

Atmospheric Gamma-ray Bursts

Currently orbiting satellites detect an average of about one gamma-ray burst per day. Because gamma-ray bursts are visible to distances encompassing most of the observable universe, a volume encompassing many billions of galaxies, this suggests that gamma-ray bursts must be exceedingly rare events per galaxy.

Measuring the exact rate of gamma-ray bursts is difficult, but for a galaxy of approximately the same size as the Milky Way, the expected rate (for long GRBs) is about one burst every 100,000 to 1,000,000 years. Only a few percent of these would be beamed towards Earth. Estimates of rates of short GRBs are even more uncertain because of the unknown beaming fraction, but are probably comparable.

A gamma-ray burst in the Milky Way, if close enough to Earth and beamed towards it, could have significant effects on the biosphere. The absorption of radiation in the atmosphere would cause photodissociation of nitrogen, generating nitric oxide that would act as a catalyst to destroy ozone. The atmospheric photodissociation:

- $N_2 \rightarrow 2N$
- $O_2 \rightarrow 2O$
- $CO_2 \rightarrow C + 2O$
- $H_2O \rightarrow 2H + O$
- $2NH_3 \rightarrow 3H_2 + N_2$

would yield,

- NO_2 (consumes up to 400 ozone molecules)
- CH_2 (nominal)
- CH_4 (nominal)
- CO_2

According to a 2004 study, a GRB at a distance of about a kiloparsec could destroy up to half of Earth's ozone layer; the direct UV irradiation from the burst combined with additional solar UV radiation passing through the diminished ozone layer could then have potentially significant impacts on the food chain and potentially trigger a mass extinction. The authors estimate that one such burst is expected per billion years, and hypothesize that the Ordovician-Silurian extinction event could have been the result of such a burst.

There are strong indications that long gamma-ray bursts preferentially or exclusively occur in regions of low metallicity. Because the Milky Way has been metal-rich since before the Earth formed, this effect may diminish or even eliminate the possibility that a long gamma-ray burst has occurred within the Milky Way within the past billion years. No such metallicity biases are known for short gamma-ray bursts. Thus, depending on their local rate and beaming properties, the possibility for a nearby event to have had a large impact on Earth at some point in geological time may still be significant.

Multiple Photon Dissociation

Single photons in the infrared spectral range usually are not energetic enough for direct photodissociation of molecules. However, after absorption of multiple infrared photons a molecule may gain internal energy to overcome its barrier for dissociation. Multiple photon dissociation (MPD, IRMPD with infrared radiation) can be achieved by applying high power lasers, e.g. a carbon dioxide laser, or a free electron laser, or by long interaction times of the molecule with the radiation field without the possibility for rapid cooling, e.g. by collisions. The latter method allows even for MPD induced by black-body radiation, a technique called blackbody infrared radiative dissociation (BIRD).

Rate Equation

The rate law or rate Equation for a chemical reaction is an equation that links the reaction rate with the concentrations or pressures of the reactants and constant parameters (normally rate coefficients and partial reaction orders). For many reactions the rate is given by a power law such as:

$$v = k[\mathrm{A}]^x[\mathrm{B}]^y$$

where [A] and [B] express the concentration of the species A and B (usually in moles per liter (molarity, M)). The exponents x and y are the partial *orders of reaction* for A and B and the *overall* reaction order is the sum of the exponents. These are often positive integers, but they may also be zero, fractional, or negative. The constant k is the reaction rate constant or *rate coefficient* of the reaction. Its value may depend on conditions such as temperature, ionic strength, surface area of an adsorbent, or light irradiation.

Elementary (single-step) reactions and reaction steps have reaction orders equal to the stoichiometric coefficients for each reactant. The overall reaction order, i.e. the sum of stoichiometric coefficients of reactants, is always equal to the molecularity of the elementary reaction. However complex (multi-step) reactions may or may not have reaction orders equal to their stoichiometric coefficients. This implies that the order and the rate equation of a given reaction cannot be reliably deduced from the stoichiometry and must be determined experimentally, since an unknown reaction mechanism could be either elementary or complex. When the experimental rate equation has been determined, it is often of use for deduction of the reaction mechanism.

The rate equation of a reaction with an assumed multi-step mechanism can often be derived theoretically using quasi-steady state assumptions from the underlying elementary reactions, and compared with the experimental rate equation as a test of the assumed mechanism. The equation may involve a fractional order, and may depend on the concentration of an intermediate species.

A reaction can also have an *undefined* reaction order with respect to a reactant if the rate is not simply proportional to some power of the concentration of that reactant; for example, one cannot talk about reaction order in the rate equation for a bimolecular reaction between adsorbed molecules:

$$v = k\frac{K_1 K_2 C_A C_B}{(1 + K_1 C_A + K_2 C_B)^2}.$$

Consider a typical chemical reaction in which two reactants A and B combine to form a product C:

$$\mathrm{A} + 2\mathrm{B} \rightarrow 3\mathrm{C}$$

This can also be written:

$$0 = -\mathrm{A} - 2\mathrm{B} + 3\mathrm{C}.$$

The prefactors -1, -2 and 3 (with negative signs for reactants because they are consumed) are known as stoichiometric coefficients. One molecule of A combines with two of B to form 3 of C, so if we use the symbol [X] for the number of moles of chemical X,

$$-\frac{d[\mathrm{A}]}{dt} = -\frac{1}{2}\frac{d[\mathrm{B}]}{dt} = \frac{1}{3}\frac{d[\mathrm{C}]}{dt}$$

If the reaction takes place in a closed system at constant temperature and volume, without a build-up of reaction intermediates, the *reaction rate r* is defined as:

$$v = \frac{1}{v_i}\frac{d[\mathrm{X}_i]}{dt},$$

where v_i is the stoichiometric coefficient for chemical X_i.

The reaction rate has some functional dependence on the concentrations of the reactants,

$$v = f([\mathrm{A}],[\mathrm{B}],\ldots),$$

and this dependence is known as the *rate equation* or *rate law*. This law generally cannot be deduced from the chemical equation and must be determined by experiment.

Power Laws

A common form for the rate equation is a power law:

$$v = k[A]^x[B]^y \ldots$$

The constant k is called the *rate constant*. The exponents, which can be fractional, are called *partial orders of reaction* and their sum is the overall order of reaction.

In a dilute solution, an elementary reaction (one having a single step with a single transition state) is empirically found to obey the law of mass action. This predicts that the rate depends only on the concentrations of the reactants, raised to the powers of their stoichiometric coefficients.

Determination of Reaction Order

Method of Initial Rates

The natural logarithm of the power-law rate equation is:

$$\ln v = \ln k + x \ln[A] + y \ln[B] + \ldots$$

This can be used to estimate the order of reaction of each reactant. For example, the initial rate can be measured in a series of experiments at different initial concentrations of reactant A with all other concentrations [B], [C], ... kept constant, so that,

$$\ln v = x \ln[A] + \text{constant}.$$

The slope of a graph of $\ln v$ as a function of $\ln[A]$ then corresponds to the order x with respect to reactant A.

However, this method is not always reliable because:

1. Measurement of the initial rate requires accurate determination of small changes in concentration in short times (compared to the reaction half-life) and is sensitive to errors, and
2. The rate equation will not be completely determined if the rate also depends on substances not present at the beginning of the reaction, such as intermediates or products.

Integral Method

The tentative rate equation determined by the method of initial rates is therefore normally verified by comparing the concentrations measured over a longer time (several half-lives) with the integrated form of the rate equation.

For example, the integrated rate law for a first-order reaction is:

$$\ln[A] = -kt + \ln[A]_0,$$

where [A] is the concentration at time t and $[A]_0$ is the initial concentration at zero time. The first-order rate law is confirmed if $\ln[A]$ is in fact a linear function of time. In this case the rate constant k is equal to the slope with sign reversed.

Method of Flooding

The partial order with respect to a given reactant can be evaluated by the method of flooding (or of isolation) of Ostwald. In this method, the concentration of one reactant is measured with all other reactants in large excess so that their concentration remains essentially constant. For a reaction a·A + b·B → c·C with rate law: $v = k \cdot [A]^x \cdot [B]^y$, the partial order x with respect to A is determined using a large excess of B. In this case:

$$v = k' \cdot [A]^x \text{wi} \;\; \text{with} \;\; k' = k \cdot [B]^y,$$

and x may be determined by the integral method. The order y with respect to B under the same conditions (with B in excess) is determined by a series of similar experiments with a range of initial concentration $[B]_0$ so that the variation of k' can be measured.

Zero Order

For zero-order reactions, the reaction rate is independent of the concentration of a reactant, so that changing its concentration has no effect on the speed of the reaction. Thus, the concentration changes linearly with time. This may occur when there is a bottleneck which limits the number of reactant molecules that can react at the same time, for example if the reaction requires contact with an enzyme or a catalytic surface.

Many enzyme-catalyzed reactions are zero order, provided that the reactant concentration is much greater than the enzyme concentration which controls the rate, so that the enzyme is *saturated*. For example, the biological oxidation of ethanol to acetaldehyde by the enzyme liver alcohol dehydrogenase (LADH) is zero order in ethanol.

Similarly reactions with heterogeneous catalysis can be zero order if the catalytic surface is saturated. For example, the decomposition of phosphine (PH_3) on a hot tungsten surface at high pressure is zero order in phosphine which decomposes at a constant rate.

In homogeneous catalysis zero order behavior can come about from reversible inhibition. For example, ring-opening metathesis polymerization using third-generation Grubbs catalyst exhibits zero order behavior in catalyst due to the reversible inhibition that is occur between the pyridine and the ruthenium center.

First Order

A *first order reaction* depends on the concentration of only one reactant (a *unimolecular reaction*). Other reactants can be present, but each will be zero order. The rate law for such a reaction is:

$$-\frac{d[A]}{dt} = k[A],$$

The half-life is independent of the starting concentration and is given by,

$$t_{1/2} = \frac{\ln(2)}{k}$$

Examples of such reactions are:

- $H_2O_2(l) \rightarrow H_2O(l) + \frac{1}{2}O_2(g)$
- $SO_2Cl_2(l) \rightarrow SO_2(g) + Cl_2(g)$
- $2N_2O_5(g) \rightarrow 4NO_2(g) + O_2(g)$

In organic chemistry, the class of SN1 (nucleophilic substitution unimolecular) reactions consists of first-order reactions. For example, in the reaction of aryldiazonium ions with nucleophiles in aqueous solution $ArN_2^+ + X^- \rightarrow ArX + N_2$, the rate equation is $v = k[ArN_2^+]$, where Ar indicates an aryl group.

Second Order

A reaction is said to be second order when the overall order is two. The rate of a second-order reaction may be proportional to one concentration squared $v = k[A]^2$, or (more commonly) to the product of two concentrations $v = k[A][B]$. As an example of the first type, the reaction $NO_2 + CO \rightarrow NO + CO_2$ is second-order in the reactant NO_2 and zero order in the reactant CO. The observed rate is given by $v = k[NO_2]^2$, and is independent of the concentration of CO.

For the rate proportional to a single concentration squared, the time dependence of the concentration is given by:

$$\frac{1}{[A]} = \frac{1}{[A]_0} + kt$$

The time dependence for a rate proportional to two unequal concentrations is:

$$\frac{[A]}{[B]} = \frac{[A]_0}{[B]_0} e^{([A]_0 - [B]_0)kt};$$

if the concentrations are equal, they satisfy the previous equation.

The second type includes nucleophillic addition-elimination reactions, such as the alkaline hydrolysis of ethyl acetate:

$$CH_3COOC_2H_5 + OH^- \rightarrow CH_3COO^- + C_2H_5OH$$

This reaction is first-order in each reactant and second-order overall: $r = k[CH_3COOC_2H_5][OH^-]$

If the same hydrolysis reaction is catalyzed by imidazole, the rate equation becomes $v = k[\text{imidazole}][CH_3COOC_2H_5]$. The rate is first-order in one reactant (ethyl acetate), and also first-order in imidazole which as a catalyst does not appear in the overall chemical equation.

Another well-known class of second-order reactions are the SN2 (bimolecular nucleophilic substitution) reactions, such as the reaction of n-butyl bromide with sodium iodide in acetone:

$$CH_3CH_2CH_2CH_2Br + NaI \rightarrow CH_3CH_2CH_2CH_2I + NaBr\downarrow$$

This same compound can be made to undergo a bimolecular (E2) elimination reaction, another common type of second-order reaction, if the sodium iodide and acetone are replaced with sodium tert-butoxide as the salt and tert-butanol as the solvent:

$$CH_3CH_2CH_2CH_2Br + NaO\textit{t}\text{-}Bu \rightarrow CH_3CH_2CH{=}CH_2 + NaBr + HO\textit{t}\text{-}Bu$$

Pseudo-first Order

If the concentration of a reactant remains constant (because it is a catalyst, or because it is in great excess with respect to the other reactants), its concentration can be included in the rate constant, obtaining a *pseudo–first-order* (or occasionally pseudo–second-order) rate equation. For a typical second-order reaction with rate equation v = k[A][B], if the concentration of reactant B is constant then v = k[A][B] = k'[A], where the pseudo–first-order rate constant k' = k[B]. The second-order rate equation has been reduced to a pseudo–first-order rate equation, which makes the treatment to obtain an integrated rate equation much easier.

One way to obtain a pseudo-first order reaction is to use a large excess of one reactant (say, [B]≫[A]) so that, as the reaction progresses, only a small fraction of the reactant in excess (B) is consumed, and its concentration can be considered to stay constant. For example, the hydrolysis of esters by dilute mineral acids follows pseudo-first order kinetics where the concentration of water is present in large excess:

$$CH_3COOCH_3 + H_2O \rightarrow CH_3COOH + CH_3OH$$

The hydrolysis of sucrose in acid solution is often cited as a first-order reaction with rate r = k[sucrose]. The true rate equation is third-order, r = k[sucrose][H^+][H_2O]; however, the concentrations of both the catalyst H^+ and the solvent H_2O are normally constant, so that the reaction is pseudo–first-order.

Reaction Orders 0, 1, 2 and *n*

Elementary reaction steps with order 3 (called *ternary reactions*) are rare and unlikely to occur. However, overall reactions composed of several elementary steps can, of course, be of any (including non-integer) order.

	Zero order	First order	Second order	*n*th order
Rate Law	$-d[A]/dt = k$	$-d[A]/dt = k[A]$	$-d[A]/dt = k[A]^2$	$-d[A]/dt = k[A]^n$
Integrated Rate Law	$[A]=[A]_0 - kt$	$[A] = [A]_0 e^{-kt}$	$\frac{1}{[A]} = \frac{1}{[A]_0} + kt$	$\frac{1}{[A]^{n-1}} \aleph \frac{1}{[A]_0^{\,n-1}}$ üü [Except first order]
Units of Rate Constant (*k*)	$\frac{M}{s}$	$\frac{1}{s}$	$\frac{1}{M \cdot s}$	$\frac{1}{M^{n-1} \cdot s}$
Linear Plot to determine *k*	[A] vs. *t*	ln([A]) vs.*t*	$\frac{1}{[A]}$ vs. *t*	$\frac{1}{[A]^{n-1}}$ vs. *t* [Except first order]
Half-life	$t_{\frac{1}{2}} = \frac{[A]_0}{2k}$	$t_{\frac{1}{2}} = \frac{\ln(2)}{k}$	$[t_{\frac{1}{2}} = \frac{1}{k[A]_0}$	$t_{\frac{1}{2}} = \lim_{x \to n} \frac{2^{x-1} - 1}{(x-1)k[A]0^{x-1}}$ [limit only necessary in first order]

Where M stands for concentration in molarity (mol · L^{-1}), t for time, and k for the reaction rate constant. The half-life of a first order reaction is often expressed as $t_{1/2} = 0.693/k$ (as ln2 = 0.693).

Fractional Order

In fractional order reactions, the order is a non-integer, which often indicates a chemical chain reaction or other complex reaction mechanism. For example, the pyrolysis of acetaldehyde (CH_3CHO) into methane and carbon monoxide proceeds with an order of 1.5 with respect to acetaldehyde: $r = k[CH_3CHO]^{3/2}$. The decomposition of phosgene ($COCl_2$) to carbon monoxide and chlorine has order 1 with respect to phosgene itself and order 0.5 with respect to chlorine: $v = k[COCl_2]\,[Cl_2]^{1/2}$.

The order of a chain reaction can be rationalized using the steady state approximation for the concentration of reactive intermediates such as free radicals. For the pyrolysis of acetaldehyde, the Rice-Herzfeld mechanism is:

Initiation:

$$CH_3CHO \rightarrow \bullet CH_3 + \bullet CHO$$

Propagation:

$$\bullet CH_3 + CH_3CHO \rightarrow CH_3CO\bullet + CH_4$$

$$CH_3CO\bullet \rightarrow \bullet CH_3 + CO$$

Termination:

$$2\ \bullet CH_3 \rightarrow C_2H_6$$

where • denotes a free radical. To simplify the theory, the reactions of the •CHO to form a second $\bullet CH_3$ are ignored.

In the steady state, the rates of formation and destruction of methyl radicals are equal, so that:

$$\frac{d[\cdot CH_3]}{dt} = k_i[CH_3CHO] - k_t[\cdot CH_3]^2 = 0,$$

so that the concentration of methyl radical satisfies:

$$[\cdot CH_3] \ \propto \ [CH_3CHO]^{\frac{1}{2}}.$$

The reaction rate equals the rate of the propagation steps which form the main reaction products CH_4 and CO:

$$v = \frac{d[CH_4]}{dt} = k_p[\cdot CH_3][CH_3CHO] \ \propto \ [CH_3CHO]^{\frac{3}{2}}$$

in agreement with the experimental order of 3/2.

Complex Laws

Mixed Order

More complex rate laws have been described as being *mixed order* if they approximate to the laws for more than one order at different concentrations of the chemical species involved. For example, a rate law of the form $v = k_1[A] + k_2[A]^2$ represents concurrent first order and second order reactions (or more often concurrent pseudo-first order and second order) reactions, and can be described as mixed first and second order. For sufficiently large values of [A] such a reaction will approximate second order kinetics, but for smaller [A] the kinetics will approximate first order (or pseudo-first order). As the reaction progresses, the reaction can change from second order to first order as reactant is consumed.

Another type of mixed-order rate law has a denominator of two or more terms, often because the identity of the rate-determining step depends on the values of the concentrations. An example is the oxidation of an alcohol to a ketone by hexacyanoferrate (III) ion [Fe(CN)6^{3-}] with ruthenate (VI) ion (RuO4^{2-}) as catalyst. For this reaction, the rate of disappearance of hexacyanoferrate (III is:

$$v = \frac{[Fe(CN)_6]^{2-}}{k_\alpha + k_\beta [Fe(CN)_6]^{2-}}$$

This is zero-order with respect to hexacyanoferrate (III) at the onset of the reaction (when its concentration is high and the ruthenium catalyst is quickly regenerated), but changes to first-order when its concentration decreases and the regeneration of catalyst becomes rate-determining.

Notable mechanisms with mixed-order rate laws with two-term denominators include:

- Michaelis-Menten kinetics for enzyme-catalysis: first-order in substrate (second-order overall) at low substrate concentrations, zero order in substrate (first-order overall) at higher substrate concentrations; and
- The Lindemann mechanism for unimolecular reactions: second-order at low pressures, first-order at high pressures.

Negative Order

A reaction rate can have a negative partial order with respect to a substance. For example, the conversion of ozone (O_3) to oxygen follows the rate equation $v = k\frac{[O_3]^2}{[O_2]}$ in an excess of oxygen. This corresponds to second order in ozone and order (-1) with respect to oxygen.

When a partial order is negative, the overall order is usually considered as undefined. In the above example for instance, the reaction is not described as first order even though the sum of the partial orders is $2 + (-1) = 1$, because the rate equation is more complex than that of a simple first-order reaction.

Opposed Reactions

A pair of forward and reverse reactions may occur simultaneously with comparable speeds. For example, A and B react into X and Y and vice versa (*s, t, u,* and *w* are the stoichiometric coefficients):

$$sA + tB \rightleftharpoons uX + wY$$

The reaction rate expression for the above reactions (assuming each one is elementary) can be expressed as:

$$v = k_1[\mathrm{A}]^s[\mathrm{B}]^t - k_2[\mathrm{X}]^u[\mathrm{Y}]^w$$

where: k_1 is the rate coefficient for the reaction that consumes A and B; k_2 is the rate coefficient for the backwards reaction, which consumes X and Y and produces A and B.

The constants k_1 and k_2 are related to the equilibrium coefficient for the reaction (K) by the following relationship (set *v*=0 in balance):

$$k_1[\mathrm{A}]^s[\mathrm{B}]^t = k_2[\mathrm{X}]^u[\mathrm{Y}]^w$$

$$K = \frac{[X]^u[Y]^w}{[A]^s[B]^t} = \frac{k_1}{k_2}$$

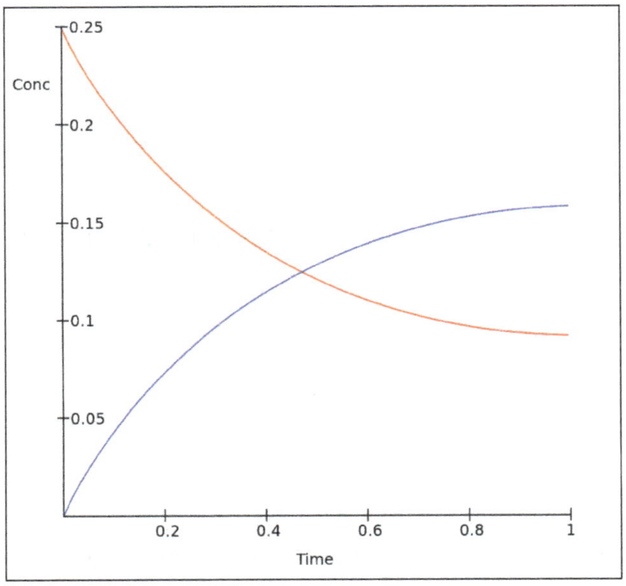

Concentration of A (A_0 = 0.25 mole/l) and B versus time reaching equilibrium k_f = 2 min^{-1} and k_r = 1 min^{-1}

Example: In a simple equilibrium between two species,

$$\mathrm{A} \rightleftharpoons \mathrm{B}$$

Where the reactions starts with an initial concentration of A, $[\mathrm{A}]_0$ with an initial concentration of 0 for B at time *t*=0.

Then the constant K at equilibrium is expressed as:

$$K \stackrel{\text{def}}{=} \frac{k_f}{k_b} = \frac{[\mathrm{B}]_e}{[\mathrm{A}]_e}$$

Where $[\mathrm{A}]_e$ and $[\mathrm{B}]_e$ are the concentrations of A and B at equilibrium, respectively.

The concentration of A at time t, $[\mathrm{A}]_t$, is related to the concentration of B at time t, $[\mathrm{B}]_t$, by the equilibrium reaction equation:

$$[\mathrm{A}]_t = [\mathrm{A}]_0 - [\mathrm{B}]_t$$

The term $[\mathrm{B}]_0$ is not present because, in this simple example, the initial concentration of B is 0.

This applies even when time t is at infinity; i.e., equilibrium has been reached:

$$[A]_e = [A]_0 - [B]_e$$

then it follows, by the definition of K, that:

$$[\mathrm{B}]_e = x = \frac{k_f}{k_f + k_b}[\mathrm{A}]_0$$

and, therefore,

$$[\mathrm{A}]_e = [\mathrm{A}]_0 - x = \frac{k_b}{k_f + k_b}[\mathrm{A}]_0$$

These equations allow us to uncouple the system of differential equations, and allow us to solve for the concentration of A alone.

The reaction equation, given previously as:

$$v = k_1[\mathrm{A}]^s[\mathrm{B}]^t - k_2[\mathrm{X}]^u[\mathrm{Y}]^v$$

$$-\frac{d[\mathrm{A}]}{dt} = k_f[\mathrm{A}]_t - k_b[\mathrm{B}]_t$$

The derivative is negative because this is the rate of the reaction going from A to B, and therefore the concentration of A is decreasing. To simplify annotation, let x be $[A]_t$, the concentration of A at time t. Let be the concentration of A at equilibrium. Then:

$$\begin{aligned}
-\frac{d[A]}{dt} &= k_f[A]_t - k_b[B]_t \\
-\frac{dx}{dt} &= k_f x - k_b[B]_t \\
&= k_f x - k_b([A]_0 - x) \\
&= (k_f + k_b)x - k_b[A]_0
\end{aligned}$$

Since:

$$k_f + k_b = k_b \frac{[A]_0}{x_e}$$

The reaction rate becomes:

$$\frac{dx}{dt} = \frac{k_b [A]_0}{x_e}(x_e - x)$$

which results in:

$$\ln\left(\frac{[A]_0 - [A]_e}{[A]_t - [A]_e}\right) = (k_f + k_b)t$$

A plot of the negative natural logarithm of the concentration of A in time minus the concentration at equilibrium versus time t gives a straight line with slope $k_f + k_b$. By measurement of A_e and B_e the values of K and the two reaction rate constants will be known.

Generalization of Simple Example

If the concentration at the time $t = 0$ is different from above, the simplifications above are invalid, and a system of differential equations must be solved. However, this system can also be solved exactly to yield the following generalized expressions:

$$[A] = [A]_0 \frac{1}{k_f + k_b}(k_b + k_f e^{-(k_f + k_b)t}) + [B]_0 \frac{k_b}{k_f + k_b}(1 - e^{-(k_f + k_b)t})$$

$$[\mathrm{B}] = [\mathrm{A}]_0 \frac{k_f}{k_f + k_b}\left(1 - e^{-(k_f + k_b)t}\right) + [\mathrm{B}]_0 \frac{1}{k_f + k_b}\left(k_f + k_b e^{-(k_f + k_b)t}\right)$$

When the equilibrium constant is close to unity and the reaction rates very fast for instance in conformational analysis of molecules, other methods are required for the determination of rate constants for instance by complete lineshape analysis in NMR spectroscopy.

Consecutive Reactions

If the rate constants for the following reaction are k_1 and k_2; $\mathrm{A} \rightarrow \mathrm{B} \rightarrow \mathrm{C}$, , then the rate equation is:

For reactant A: $\dfrac{d[A]}{dt} = -k_1[A]$

For reactant B: $\dfrac{d[B]}{dt} = k_1[A] - k_2[B]$

For product C: $\dfrac{d[C]}{dt} = k_2[B]$

With the individual concentrations scaled by the total population of reactants to become probabilities, linear systems of differential equations such as these can be formulated as a master equation. The differential equations can be solved analytically and the integrated rate equations are:

$$[A]=[A]_0 e^{-k_1 t}$$

$$[\mathrm{B}]=\begin{cases}[A]_0 \dfrac{k_1}{k_2-k_1}(e^{-k_1 t}-e^{-k_2 t})+[B]_0 e^{-k_2 t} & k_1 \neq k_2 \\ [A]_0 k_1 t e^{-k_1 t}+[B]_0 e^{-k_1 t} & \text{otherwise}\end{cases}$$

$$[C]=\begin{cases}[A]_0\left(1+\dfrac{k_1 e^{-k_2 t}-k_2 e^{-k_1 t}}{k_2-k_1}\right)+[B]_0\left(1-e^{-k_2 t}\right)+[C]_0 & k_1 \neq k_2 \\ [A]_0\left(1-e^{-k_1 t}-k_1 t e^{-k_1 t}\right)+[B]_0\left(1-e^{-k_1 t}\right)+[C]_0 & \text{otherwise}\end{cases}$$

The steady state approximation leads to very similar results in an easier way.

Parallel or Competitive Reactions

Time course of two first order, competitive reactions with differing rate constants.

Two first order, competitive reactions

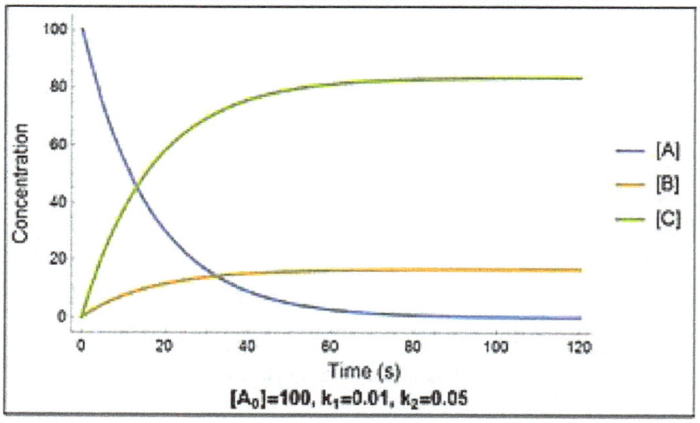

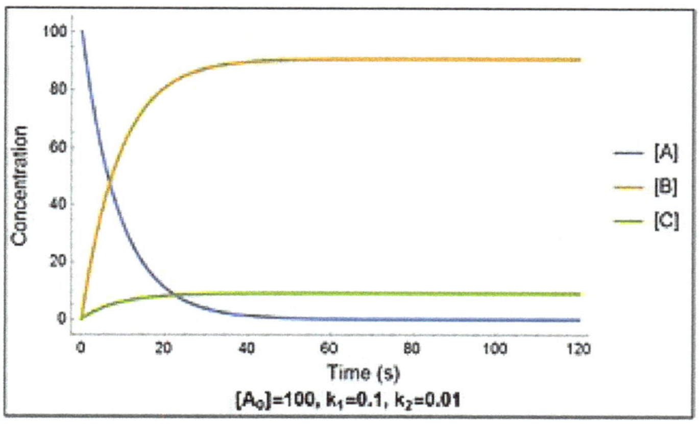

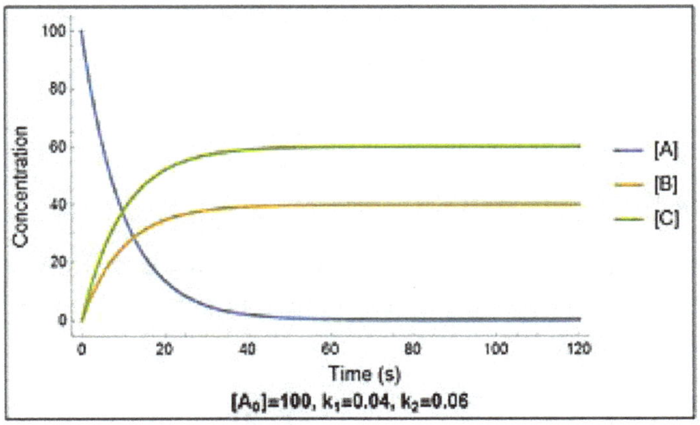

When a substance reacts simultaneously to give two different products, a parallel or competitive reaction is said to take place.

Two First Order Reactions

$A \longrightarrow B$ and $A \longrightarrow C$, with constants k_1 and k_2 and rate equations $-\dfrac{d[A]}{dt} = (k_1 + k_2)[A]$; $\dfrac{d[B]}{dt} = k_1[A]$ and $\dfrac{d[C]}{dt} = k_2[A]$.

The integrated rate equations are then $[A] = [A]_0 e^{-(k_1+k_2)t}$; $[B] = \dfrac{k_1}{k_1+k_2}[A]_0(1 - e^{-(k_1+k_2)t})$ and $[C] = \dfrac{k_2}{k_1+k_2}[A]_0(1 - e^{-(k_1+k_2)t})$.

One important relationship in this case is $\dfrac{[B]}{[C]} = \dfrac{k_1}{k_2}$.

One First Order and One Second Order Reaction

This can be the case when studying a bimolecular reaction and a simultaneous hydrolysis (which can be treated as pseudo order one) takes place: the hydrolysis complicates the study of the reaction kinetics, because some reactant is being "spent" in a parallel reaction. For example, A reacts with R to give our product C, but meanwhile the hydrolysis reaction takes away an amount of A to give B, a by product: $\mathrm{A} + \mathrm{H_2O} \longrightarrow \mathrm{B}$ and ü $\rightarrow$ C. The rate equations are: $\dfrac{d[B]}{dt} = k_1[\mathrm{A}][\mathrm{H_2O}] = k_1'[\mathrm{A}]$ and $\dfrac{d[C]}{dt} = k_2[A][R]$. Where k_1' is the pseudo first order constant.

The integrated rate equation for the main product [C] is $[C] = [R]_0 \left[1 - e^{-\frac{k_2}{k_1'}[A]_0(1-e^{-k_1' t})} \right]$, which is equivalent to $\ln \dfrac{[R]_0}{[R]_0 - [C]} = \dfrac{k_2[A]_0}{k_1'}(1 - e^{-k_1' t})$. Concentration of B is related to that of C through

$$[B] = -\frac{k_1'}{k_2} ln \left(1 - \frac{[C]}{[R]_0} \right)$$

The integrated equations were analytically obtained but during the process it was assumed that $[A]_0 - [C] \approx [A]_0$ therefeore, previous equation for [C] can only be used for low concentrations of [C] compared to $[A]_0$.

Stoichiometric Reaction Networks

The most general description of a chemical reaction network considers a number *N* of distinct chemical species reacting via *R* reactions. The chemical equation of the *j*-th reaction can then be written in the generic form:

$$s_{1j}X_1 + s_{2j}X_2 \ldots + s_{Nj}X_N \xrightarrow{k_j} r_{1j}X_1 + r_{2j}X_2 + \ldots + r_{Nj}X_N,$$

which is often written in the equivalent form:

$$\sum_{i=1}^{N} s_{ij}X_i \xrightarrow{k_j} \sum_{i=1}^{N} r_{ij}X_i.$$

Here,

- *j* is the reaction index running from 1 to *R*,
- X_i denotes the *i*-th chemical species,
- k_j is the rate constant of the j-th reaction and
- s_{ij} and r_{ij} are the stoichiometric coefficients of reactants and products, respectively.

The rate of such reaction can be inferred by the law of mass action:

$$f_j([\vec{X}]) = k_j \prod_{z=1}^{N} [X_z]^{s_{zj}}$$

which denotes the flux of molecules per unit time and unit volume. Here $[\vec{X}] = ([X1],[X2],\ldots,[X_N])$ is the vector of concentrations. This definition includes the elementary reactions:

- Zero order reactions:

  for which $s_{zj} = 0$ for all *z*,

- First order reactions:

  for which $s_{zj} = 1$ for a single *z*,

- Second order reactions:

  for which $s_{zj} = 1$ for exactly two *z*, i.e., a bimolecular reaction, or $s_{zj} = 2$ for a single *z*, i.e., a dimerization reaction.

One can define the stoichiometric matrix:

$$S_{ij} = r_{ij} - s_{ij},$$

denoting the net extent of molecules of i in reaction j. The reaction rate equations can then be written in the general form:

$$\frac{d[\mathrm{X}_i]}{dt} = \sum_{j=1}^{R} S_{ij} f_j([\vec{\mathrm{X}}]).$$

This is the product of the stoichiometric matrix and the vector of reaction rate functions. Particular simple solutions exist in equilibrium, $\frac{d[\mathrm{X}_i]}{dt} = 0$, for systems composed of merely reversible reactions. In this case the rate of the forward and backward reactions are equal, a principle called detailed balance. Detailed balance is a property of the stoichiometric matrix S_{ij} alone and does not depend on the particular form of the rate functions f_j. All other cases where detailed balance is violated are commonly studied by flux balance analysis which has been developed to understand metabolic pathways.

General Dynamics of Unimolecular Conversion

For a general unimolecular reaction involving interconversion of N different species, whose concentrations at time t are denoted by $X_1(t)$ through $X_N(t)$, an analytic form for the time-evolution of the species can be found. Let the rate constant of conversion from species X_i to species X_j be denoted as k_{ij}., and construct a rate-constant matrix K whose entries are the k_{ij}.

Also, let $X(t) = (X_1(t), X_2(t), ..., X_N(t))^T$ be the vector of concentrations as a function of time.

Let $J = (1,1,1,...,1)^T$ be the vector of ones.

Let I be the $N \times N$ identity matrix.

Let Diag be the function that takes a vector and constructs a diagonal matrix whose on-diagonal entries are those of the vector.

Let $\mathcal{L}^{-1}$ be the inverse Laplace transform from s to t.

Then the time-evolved state $X(t)$ is given by:

$$X(t) = \mathcal{L}^{-1}[(sI + Diag(KJ) - K^T)^{-1} X(0)],$$

thus providing the relation between the initial conditions of the system and its state at time t.

Chemical Reaction Rates

A *rate* is a measure of how some property varies with time. Speed is a familiar rate that expresses the distance traveled by an object in a given amount of time. Wage is a rate that represents the amount of money earned by a person working for a given amount of time. Likewise, the rate of a chemical reaction is a measure of how much reactant is consumed, or how much product is produced, by the reaction in a given amount of time.

The rate of reaction is the change in the amount of a reactant or product per unit time. Reaction rates are therefore determined by measuring the time dependence of some property that can be related to reactant or product amounts. Rates of reactions that consume or produce gaseous substances, for example, are conveniently determined by measuring changes in volume or pressure. For reactions involving one or more colored substances, rates may be monitored via measurements of light absorption. For reactions involving aqueous electrolytes, rates may be measured via changes in a solution's conductivity.

For reactants and products in solution, their relative amounts (concentrations) are conveniently used for purposes of expressing reaction rates. If we measure the concentration of hydrogen peroxide, H_2O_2, in an aqueous solution, we find that it changes slowly over time as the H_2O_2 decomposes, according to the equation:

$$2H_2O_2(aq) \rightarrow 2H_2O(l) + O_2(g)$$

The rate at which the hydrogen peroxide decomposes can be expressed in terms of the rate of change of its concentration, as shown here:

$$\text{rate of decomposition of } H_2O_2 = -\frac{\text{change in cocentration of reactant}}{\text{time interval}} = -\frac{[H_2O_2]t_2 - [H_2O_2]t_1}{t_2 - t_1} = -\frac{\Delta[H_2O_2]}{\Delta t}$$

This mathematical representation of the change in species concentration over time is the rate expression for the reaction. The brackets indicate molar concentrations, and the symbol delta (Δ) indicates "change in." Thus, $[H_2O_2]_{t_1}$ represents the molar concentration of hydrogen peroxide at some time t_1; likewise, $[H_2O_2]_{t_2}$ represents the molar concentration of hydrogen peroxide at a later time t_2; and $\Delta[H_2O_2]$ represents the change in molar concentration of hydrogen peroxide during the time interval Δt (that is, $t_2 - t_1$). Since the reactant concentration decreases as the reaction proceeds, $\Delta[H_2O_2]$ is a negative quantity; we place a negative sign in front of the expression because reaction rates are, by convention, positive quantities. Figure provides an example of data collected during the decomposition of H_2O_2.

Time (h)	$[H_2O_2]$ (mol L^{-1})	$\Delta[H_2O_2]$ (mol L^{-1})	Δt (h)	Rate of Decomposition, (mol/L/h)
0.00	1.000			
		−0.500	6.00	−0.0833
6.00	0.500			
		−0.250	6.00	−0.0417
12.00	0.250			
		−0.125	6.00	−0.0208
18.00	0.125			
		−0.062	6.00	−0.0103
24.00	0.0625			

The rate of decomposition of H_2O_2 in an aqueous solution decreases as the concentration of H_2O_2 decreases.

To obtain the tabulated results for this decomposition, the concentration of hydrogen peroxide was measured every 6 hours over the course of a day at a constant temperature of 40 °C. Reaction rates were computed for each time interval by dividing the change in concentration by the corresponding time increment, as shown here for the first 6-hour period:

$$\frac{-\Delta[H_2O_2]}{\Delta t}=\frac{-(0.500\text{mol/L}-1.000\text{mol/L})}{(6.00\,\text{h}-0.00\,\text{h})}=0.0833\,\text{molL}^{-1}\,\text{h}^{-1}$$

Notice that the reaction rates vary with time, decreasing as the reaction proceeds. Results for the last 6-hour period yield a reaction rate of:

$$\frac{-\Delta[H_2O_2]}{\Delta t}=\frac{-(0.0625\text{mol/L}-0.125\text{mol/L})}{(24.00\,\text{h}-18.00\,\text{h})}=0.0104\,\text{molL}^{-1}\,\text{h}^{-1}$$

This behavior indicates the reaction continually slows with time. Using the concentrations at the beginning and end of a time period over which the reaction rate is changing results in the calculation of an average rate for the reaction over this time interval. At any specific time, the rate at which a reaction is proceeding is known as its instantaneous rate. The instantaneous rate of a reaction at "time zero," when the reaction commences, is its initial rate. Consider the analogy of a car slowing down as it approaches a stop sign. The vehicle's initial rate—analogous to the beginning of a chemical reaction—would be the speedometer reading at the moment the driver begins pressing the brakes (t_0). A few moments later, the instantaneous rate at a specific moment—call it t_1—would be somewhat slower, as indicated by the speedometer reading at that point in time. As time passes, the instantaneous rate will continue to fall until it reaches zero, when the car (or reaction) stops. Unlike instantaneous speed, the car's average speed is not indicated by the speedometer; but it can be calculated as the ratio of the distance traveled to the time required to bring the vehicle to a complete stop (Δt). Like the decelerating car, the average rate of a chemical reaction will fall somewhere between its initial and final rates.

The instantaneous rate of a reaction may be determined one of two ways. If experimental conditions permit the measurement of concentration changes over very short time intervals, then average rates computed provide reasonably good approximations of instantaneous rates. Alternatively, a graphical procedure may be used that, in effect, yields the results that would be obtained if short time interval measurements were possible. If we plot the concentration of hydrogen peroxide against time, the instantaneous rate of decomposition of H_2O_2 at any time t is given by the slope of a straight line that is tangent to the curve at that time.

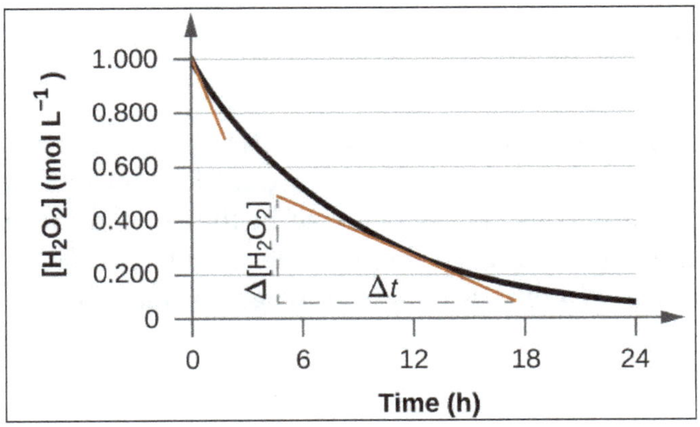

This graph shows a plot of concentration versus time for a 1.000 M solution of H_2O_2. The rate at any instant is equal to the opposite of the slope of a line tangential to this curve at that time. Tangents are shown at t = 0 h ("initial rate") and at t = 10 h ("instantaneous rate" at that particular time).

Relative Rates of Reaction

The rate of a reaction may be expressed in terms of the change in the amount of any reactant or product, and may be simply derived from the stoichiometry of the reaction. Consider the reaction represented by the following equation:

$$2\mathrm{NH_3}(g) \rightarrow \mathrm{N_2}(g) + 3\mathrm{H_2}(g)$$

The stoichiometric factors derived from this equation may be used to relate reaction rates in the same manner that they are used to related reactant and product amounts. The relation between the reaction rates expressed in terms of nitrogen production and ammonia consumption, for example, is:

$$-\frac{\Delta \mathrm{mol NH_3}}{\Delta t} \times \frac{1\mathrm{mol N_2}}{2\mathrm{mol NH_3}} = \frac{\Delta \mathrm{mol N_2}}{\Delta t}$$

We can express this more simply without showing the stoichiometric factor's units:

$$-\frac{1}{2}\frac{\Delta \mathrm{mol NH_3}}{\Delta t} = \frac{\Delta \mathrm{mol NH_2}}{\Delta t}$$

Note that a negative sign has been added to account for the opposite signs of the two amount changes (the reactant amount is decreasing while the product amount is increasing). If the reactants and products are present in the same solution, the molar amounts may be replaced by concentrations:

$$-\frac{1}{2}\frac{\Delta[\mathrm{NH_3}]}{\Delta t} = \frac{\Delta[\mathrm{N_2}]}{\Delta t}$$

Similarly, the rate of formation of H_2 is three times the rate of formation of N_2 because three moles of H_2 form during the time required for the formation of one mole of N_2:

$$\frac{1}{3}\frac{\Delta[\mathrm{H_2}]}{\Delta t} = \frac{\Delta[\mathrm{N_2}]}{\Delta t}$$

Figure illustrates the change in concentrations over time for the decomposition of ammonia into nitrogen and hydrogen at 1100 °C. We can see from the slopes of the tangents drawn at t = 500 seconds that the instantaneous rates of change in the concentrations of the reactants and products are related by their stoichiometric factors. The rate of hydrogen production, for example, is observed to be three times greater than that for nitrogen production:

$$\frac{2.90\times10^{-6} M/s}{9.71\times10^{-6} M/s} \approx 3$$

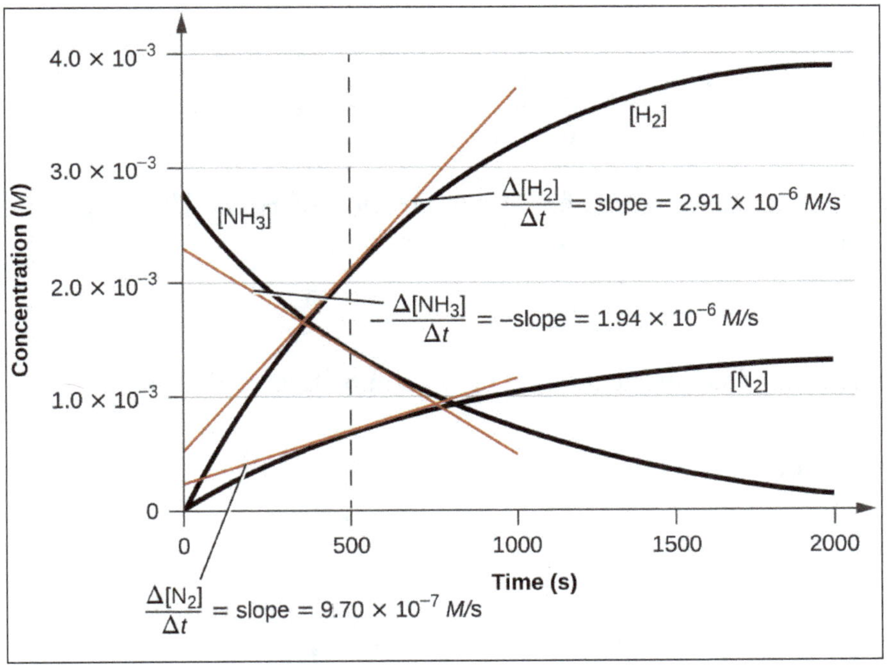

This graph shows the changes in concentrations of the reactants and products during the reaction $2NH_3 \rightarrow 3N_2 + H_2$. The rates of change of the three concentrations are related by their stoichiometric factors, as shown by the different slopes of the tangents at $t = 500$ s.

Example:

Expressions for Relative Reaction Rates

The first step in the production of nitric acid is the combustion of ammonia:

$$4NH_3(g) + 5O_2(g) \rightarrow 4NO(g) + 6H_2O(g)$$

Write the equations that relate the rates of consumption of the reactants and the rates of formation of the products.

Solution: Considering the stoichiometry of this homogeneous reaction, the rates for the consumption of reactants and formation of products are:

$$-\frac{1}{4}\frac{\Delta[NH_3]}{\Delta t} = -\frac{1}{5}\frac{\Delta[O_2]}{\Delta t} = -\frac{1}{4}\frac{\Delta[NO]}{\Delta t} = \frac{1}{6}\frac{\Delta[H_2O]}{\Delta t}$$

Example: The rate of formation of Br_2 is 6.0×10^{-6} mol/L/s in a reaction described by the following net ionic equation:

$$5Br^- + BrO_3^- + 6H^+ \rightarrow 3Br_2 + 3H_2O$$

The equations that relate the rates of consumption of the reactants and the rates of formation of the products.

Solution:

$$-\frac{1}{5}\frac{\Delta\left[\mathrm{Br}^-\right]}{\Delta t}=-\frac{\Delta\left[\mathrm{BrO_3}^-\right]}{\Delta t}=-\frac{1}{6}\frac{\Delta\left[\mathrm{H}^+\right]}{\Delta t}=\frac{1}{3}\frac{\Delta[\mathrm{Br_2}]}{\Delta t}=\frac{1}{3}\frac{\Delta[\mathrm{H_2O}]}{\Delta t}$$

Example: Reaction Rate Expressions for Decomposition of H_2O_2. The graph in previous figure shows the rate of the decomposition of H_2O_2 over time:

$$2\mathrm{H_2O_2} \rightarrow 2\mathrm{H_2O}+\mathrm{O_2}$$

Based on these data, the instantaneous rate of decomposition of H_2O_2 at t = 11.1 h is determined to be:

3.20×10^{-2} mol/L/h, that is:

$$-\frac{\Delta(\mathrm{H_2O_2})}{\Delta t}=3.20\times10^{-2}\,\mathrm{mol\,L^{-1}\,h^{-1}}$$

What is the instantaneous rate of production of H_2O and O_2?

Solution: Using the stoichiometry of the reaction, we may determine that:

$$-\frac{1}{2}\frac{\Delta(\mathrm{H_2O_2})}{\Delta t}=\frac{1}{2}\frac{\Delta(\mathrm{H_2O})}{\Delta t}=\frac{\Delta(O_2)}{\Delta t}$$

Therefore:

$$\frac{1}{2}\times3.20\times10^{-2}\,\mathrm{mol\,L^{-1}\,h^{-1}}=\frac{\Delta(O_2)}{\Delta t}$$

and

$$\frac{\Delta(O_2)}{\Delta t}=1.60\times10^{-2}\,\mathrm{mol\,L^{-1}\,h^{-1}}$$

Example: If the rate of decomposition of ammonia, NH_3, at 1150 K is 2.10×10^{-6} mol/L/s, what is the rate of production of nitrogen and hydrogen?

Solution: 1.05×10^{-6} mol/L/s, N_2 and 3.15×10^{-6} mol/L/s, H_2.

Chemical Kinetics

Chemical kinetics is the branch of chemistry that is concerned with understanding the rates of chemical reactions. It is to be contrasted with thermodynamics, which deals with the direction in which a process occurs but in itself tells nothing about its rate. Thermodynamics is time's arrow, while chemical kinetics is time's clock. Chemical kinetics relates to many aspects of cosmology, geology, biology, engineering, and even psychology and thus has far-reaching implications. The principles of chemical kinetics apply to purely physical processes as well as to chemical reactions.

One reason for the importance of kinetics is that it provides evidence for the mechanisms of chemical processes. Besides being of intrinsic scientific interest, knowledge of reaction mechanisms is of practical use in deciding what is the most effective way of causing a reaction to occur. Many commercial processes can take place by alternative reaction paths, and knowledge of the mechanisms makes it possible to choose reaction conditions that favour one path over others.

A chemical reaction is, by definition, one in which chemical substances are transformed into other substances, which means that chemical bonds are broken and formed so that there are changes in the relative positions of atoms in molecules. At the same time, there are shifts in the arrangements of the electrons that form the chemical bonds. A description of a reaction mechanism must therefore deal with the movements and speeds of atoms and electrons. The detailed mechanism by which a chemical process occurs is referred to as the reaction path, or pathway.

The vast amount of work done in chemical kinetics has led to the conclusion that some chemical reactions go in a single step; these are known as elementary reactions. Other reactions go in more than one step and are said to be stepwise, composite, or complex. Measurements of the rates of chemical reactions over a range of conditions can show whether a reaction proceeds by one or more steps. If a reaction is stepwise, kinetic measurements provide evidence for the mechanism of the individual elementary steps. Information about reaction mechanisms is also provided by certain nonkinetic studies, but little can be known about a mechanism until its kinetics has been investigated. Even then, some doubt must always remain about a reaction mechanism. An investigation, kinetic or otherwise, can disprove a mechanism but can never establish it with absolute certainty.

Reaction Rate

The rate of a reaction is defined in terms of the rates with which the products are formed and the reactants (the reacting substances) are consumed. For chemical systems it is usual to deal with the concentrations of substances, which is defined as the amount of substance per unit volume. The rate can then be defined as the concentration of a substance that is consumed or produced in unit time. Sometimes it is more convenient to express rates as numbers of molecules formed or consumed in unit time.

The Half-life

A useful rate measure is the half-life of a reactant, which is defined as the time that it takes for half of the initial amount to undergo reaction. For a special type of kinetic behaviour the half-life is independent of the initial amount. A common and straightforward example of a half-life independent of the initial amount is radioactive substances. Uranium-238, for example, decays with a half-life of 4.5 billion years; of an initial amount of uranium, half of that amount will have decayed in that period of time. The same behaviour is found in many chemical reactions.

Even when the half-life of a reaction varies with the initial conditions, it is often convenient to quote a half-life, bearing in mind that it applies only to the particular initial conditions. Consider, for example, the reaction in which hydrogen and oxygen gases combine to form water; the chemical equation is:

$$2H_2 + O_2 \rightarrow 2H_2O.$$

If the gases are mixed together at atmospheric pressure and room temperature, nothing observable will happen over long periods of time. However, reaction does occur, with a half-life that is estimated to be more than 12 billion years, which is roughly the age of the universe. If a spark is passed through the system, the reaction occurs with explosive violence, with a half-life of less than one-millionth of a second. This is a striking example of the great range of rates with which chemical kinetics is concerned. There are many possible processes that proceed too slowly to be studied experimentally, but sometimes they can be accelerated, often by the addition of a substance known as a catalyst. Some reactions are even faster than the hydrogen-oxygen explosion—for example, the combination of atoms or molecular fragments (called free radicals) where all that occurs is the formation of a chemical bond. Some modern kinetic investigations are concerned with even faster processes, such as the breakdown of highly energetic and therefore transient molecules, where times of the order of femtoseconds (fs; 1 fs = 10^{-15} second) are involved.

Measuring Slow Reactions

The best way to study exceedingly slow reactions is to change the conditions so that the reactions occur in a reasonable time. Increasing the temperature, which can have a strong effect on the reaction rate, is one possibility. If the temperature of a hydrogen-oxygen mixture is raised to about 500 °C (900 °F), reaction then occurs rapidly, and its kinetics has been studied under those conditions. When a reaction occurs to a measurable extent over a period of minutes, hours, or days, rate measurements are straightforward. Amounts of reactants or products are measured at various times, and the rates are readily calculated from the results. Many automated systems have now been devised for measuring rates in this way.

Measuring Fast Reactions

Some processes are so fast that special techniques have to be used to study them. There are two difficulties with fast reactions. One is that the time that it takes to mix reactants or to change the temperature of the system may be significant in comparison with the half-life, so that the initial time cannot be measured accurately. The other difficulty is that the time it takes to measure the amounts of substances may be comparable with the half-life of the reaction. The methods used to overcome these difficulties fall into two classes: flow methods and pulse and probe methods.

In flow methods, two gases or solutions are introduced rapidly into a mixing vessel, and the resulting mixture then flows rapidly along a tube. Concentrations of reactants or products may then be measured—for example, by spectroscopic methods—at various positions along the tube, which correspond to various reaction times. A modification of this method is the stopped-flow technique, in which the reactants are forced rapidly into a reaction chamber; the flow is then suddenly stopped, and the amounts are measured by physical methods after various short times. These flow methods are limited by the time it takes to mix gases or solutions and are not suitable if the half-life is less than about a hundredth of a second.

These mixing difficulties were overcome by pulse and probe methods. The principle of these is that a short pulse, usually of radiation, is given to a chemical system and is then followed by a probe, usually involving radiation that provides spectroscopic evidence of what occurred after the initial pulse. The first of these methods, developed in 1949 by British chemists R.G.W. Norrish and George Porter, was the flash-photolysis method, for which Norrish and Porter won the Nobel Prize for Chemistry in 1967. In this technique a flash of light of high intensity but short duration brings about the formation of atomic and molecular species, the reactions of which can be studied kinetically by spectroscopy. In the earliest experiments the duration of the flash was about a millisecond (ms; 1 ms = 10^{-3} second), but in the next four decades the duration was reduced by more than 11 powers of 10, to just a few femtoseconds. A nanosecond (ns; 1 ns = 10^{-9} second) flash is adequate for studying almost any purely chemical reaction where there is a change in chemical identity. Any chemical reaction, however, involves processes of a purely physical nature, such as energy redistribution and the breakdown of transient species, which occur in the femtosecond range.

Many such processes have now been studied with flashes of only a few femtoseconds' duration. The time that it takes for the length of a chemical bond to change by 10^{-10} metre can be as little as about 100 fs, so that a flash of a few femtoseconds' duration, closely followed by another one of the same duration, will provide information about such tiny changes in bond lengths. The technique for causing one flash to occur a few nanoseconds after another is to route the light by a slightly longer path. A path of 1 additional micrometre (μm; 1 μm = 10^{-6} metre) causes a delay of 1 fs, and such a short path difference is now technically feasible. Egyptian-born chemist Ahmed Zewail won the Nobel Prize for Chemistry in 1999 for his work in this field.

Another pulse method is the relaxation method, developed in the 1950s by German physicist Manfred Eigen. In this method the investigation begins with a reaction system in equilibrium; the reaction to be studied has finished, and no further changes take place. The external conditions are then altered very rapidly; the system is then no longer at equilibrium, and it relaxes to a new equilibrium. The speed of relaxation is measured by a physical method such as spectroscopy, and analysis of the results leads to the reaction rate.

The most common way of changing the external conditions is to change the temperature, and the method is called the temperature-jump, or T-jump, method. Techniques have been developed for raising the temperature of a tiny reaction vessel by a few degrees in less than 100 ns. The method is therefore not suitable for the fastest processes, which can be studied by flash photolysis, but many purely chemical processes are suitable for the T-jump technique, which has provided valuable kinetic information.

Other experimental techniques are used for the study of rapid processes. Ultrasonic methods have been used for processes occurring with half-lives in the microsecond (μs; 1 μs = 10^{-6} second) and nanosecond ranges. Nuclear magnetic resonance has also been used for certain types of reactions.

Some Kinetic Principles

The kinetic behaviour of an ordinary chemical reaction is conventionally studied in the first instance by determining how the reaction rate is influenced by certain external factors such as the concentrations of the reacting substances, the temperature, and sometimes the pressure. For a reaction in which two substances A and B react with each other, it is sometimes found that the reaction rate is proportional to the concentration of A, represented by [A], and to the concentration of B, or [B]. In that case the reaction is said to be a second-order reaction; it is first order in [A] and first order in [B]. In such a case the reaction rate v can be expressed as:

$$v = k[A][B],$$

where k is a constant, known as the rate constant for the reaction.

This is just one of many types of kinetics that can be observed. A substance A that changes into another substance may obey a kinetic equation of the form $v = k[A]$, which is a first-order reaction. It is important to recognize that the kinetics of a reaction does not always correspond in a simple way to the balanced chemical equation for the reaction. Thus, if a reaction is of the form:

$$A + B \rightleftharpoons Y + Z,$$

the reaction is not necessarily second-order in both directions. This is in contrast to the situation with the equilibrium constant for the reaction, which corresponds to the balanced equation. The reason why the kinetic law is different is that the reactions in the forward and reverse directions may occur by stepwise mechanisms that lead to a different and usually more complex kinetic equation.

Sometimes reaction rates depend on reactant concentrations in a more complicated way. This is a clear indication that a reaction happens in several steps.

The effect of temperature on reaction rates provides much information about reaction mechanisms. Understanding of this effect owes much to the ideas of the Dutch physical chemist Jacobus Henricus van 't Hoff and the Swedish chemist Svante August Arrhenius. Their equation for the dependence of a rate constant k on the absolute temperature T is:

$$k = A \exp(-E/RT),$$

where R is the molar gas constant and A and E are quantities that are different for each reaction. This equation has come to be called the Arrhenius equation, although, as Arrhenius acknowledged when he applied it in 1889, it was first suggested by van 't Hoff in 1884. According to this relationship, a plot of the logarithm of the rate constant against the reciprocal of the absolute temperature should yield a straight line. From the slope and intercepts of the line, it is possible to calculate the value of the kinetic parameters A and E. The Arrhenius relationship applies satisfactorily to most reactions and indeed to many physical processes; however, various complications may cause it to fail.

If the reaction between two molecules is an elementary one, occurring in a single step, a simple interpretation of the Arrhenius equation can be given. The quantity A is related to the frequency of collisions between the reacting molecules. The quantity E, known as the activation energy for the reaction, results from the fact that there is an energy barrier to reaction. If E was zero, k would be equal to A, which means that the reaction would occur every time a collision occurred between the reactant molecules. This is the case for reactions in which no chemical bond is broken, such as the combination of atoms.

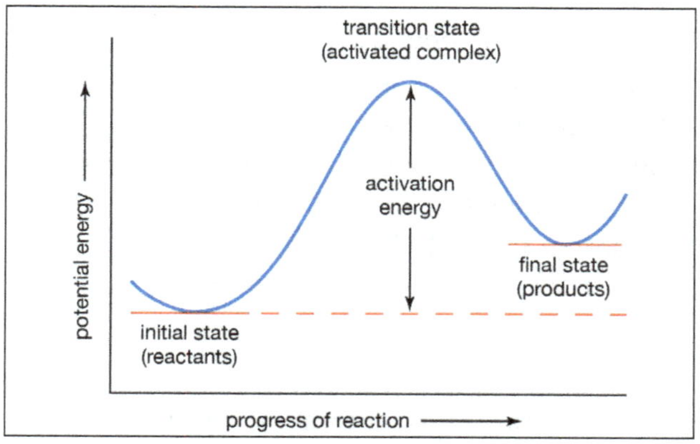

In the Arrhenius equation, the activation energy (E) represents the minimum amount of energy required to transform reactants into products in a chemical reaction. On a potential energy curve, the value of the activation energy is equivalent to the difference in potential energy between particles in an intermediate configuration (known as the activated complex, or transition state) and particles of reactants in their initial state. The activation energy thus can be visualized as a barrier that must be overcome by reactants before products can be formed.

For reactions in which a chemical bond is broken, on the other hand, the activation energy E is not zero but has a value that is often a tenth or so of the energy required to break the bond. A simple and essentially correct explanation of the activation energy was suggested by Arrhenius, who pointed out that, for many reactions, raising the temperature by 10 °C (18 °F) doubles the reaction rate. This increase cannot be caused by the increase in the frequency of collisions between colliding molecules, since the frequency does not increase sufficiently with a rise in temperature. Arrhenius suggested that when reactants A and B react together, they first form a highly energized intermediate that is denoted as $AB^‡$, which subsequently gives the products of reaction:

$$A + B \rightleftharpoons AB^‡ \rightarrow Y + Z.$$

If the intermediate complex (also called the activated complex) $AB^‡$ is of high energy, it is formed only in small amounts. According to the Boltzmann principle, the fraction of molecules having energy greater than E is $\exp(-E/RT)$, which provides an explanation of the appearance of this fraction in the Arrhenius equation. The interpretation of the equation is thus that only those molecules having energy greater than E are able to undergo reaction; other collisions are ineffective, and the reactant molecules merely separate unchanged.

Composite Reaction Mechanisms

Various lines of evidence are used to determine if a reaction occurs in more than one step. Suppose that the kinetic equation for the reaction does not correspond to the balanced equation for the reaction. A simple example is the reaction between hydrogen and iodine chloride, with the formation of iodine and hydrogen chloride:

$$H_2 + 2ICl \rightarrow I_2 + 2HCl.$$

To make the equation balance, the reaction must be written as shown, with two iodine chloride molecules reacting with a single hydrogen molecule. If this reaction occurred in a single elementary step, the rate would be proportional to the first power of the hydrogen concentration and the square of the iodine chloride concentration. Instead, however, the rate is found to be proportional to both concentrations to the first power, so that it is a second-order reaction:

$$v = k[H_2][ICl].$$

This can be explained if there is initially a slow reaction between one hydrogen molecule and one of iodine chloride:

$$H_2 + ICl \rightarrow HI + HCl \text{ (slow)}$$

followed by a rapid reaction between the hydrogen iodide formed and an additional molecule of iodine chloride:

$$HI + ICl \rightarrow HCl + I_2 \text{ (fast).}$$

If the second reaction is fast, the hydrogen iodide is removed as fast as it is formed. The rate of the second reaction therefore has no effect on the overall rate, which is the rate of the first step. This mechanism therefore explains the kinetic behaviour but does not prove it; other, more complicated schemes could be devised, but, until there is further evidence, it is expedient to accept the simple mechanism. This is an example of a consecutive reaction, which occurs in two steps, with the intermediate playing a role.

Another piece of evidence for a composite mechanism is the detection of reaction intermediates. In such a case, a reaction scheme must be devised that will account for these intermediates. Sometimes an intermediate can be a fairly stable substance. In other cases the intermediates are unstable species such as atoms and free radicals (fragments of molecules) that subsequently undergo rapid reactions. Free radicals can be detected by spectroscopy and other means. When organic molecules are raised to high temperatures, they decompose into smaller molecules, and organic free radicals have often been detected as intermediates. In an explosion, such as that between hydrogen and oxygen, free radicals such as hydroxyl can be detected.

Composite reaction mechanisms are of various kinds. Aside from the simple consecutive schemes, there are some special mechanisms that give rise to oscillatory behaviour: the amount of a product continuously rises and falls over a period of time. The conditions for this behaviour are that there must be at least two species involved in the reaction and there must be feedback, which means that products of the reaction affect the rate. There are also reaction mechanisms that give rise to what is technically known as chaos, or catastrophe. With such reactions it is impossible to predict the outcome. Chaotic conditions also require that there be feedback and that at least three species be involved.

Sometimes a complex reaction mechanism involves a cycle of reactions such that certain intermediates consumed in one step are regenerated in another. For example, the accepted mechanism of the reaction between hydrogen and bromine, which can be written as:

$$H_2 + Br_2 \rightarrow 2HBr,$$

includes the steps:

$$Br + H_2 \rightarrow HBr + H$$

$$H + Br_2 \rightarrow HBr + Br.$$

In the first of these steps a bromine atom is consumed, but in the second a bromine atom is regenerated. This pair of reactions can thus occur with the production of two molecules of hydrogen bromide, the product of the reaction, without loss of bromine atoms. This pair of reactions is called a cycle of reactions, and it can occur a number of times, in which case the reaction is referred to as a chain reaction. The two reactions in which bromine is regenerated are known as the chain-propagating steps. The average number of times the pair of steps is repeated is known as the chain length.

One necessary condition for a proposed reaction mechanism to be correct is that it must account for the overall kinetic behaviour of the reaction—in particular, for the dependence of the reaction rate on the reactant concentrations. For any proposed reaction mechanism, it is possible to write down equations for the rate of each step in terms of the reactant concentration and then to solve the equations for the overall rate. A practical difficulty arises, since no exact mathematical solution is possible for all except the simplest of mechanisms. If one has values for the rate constants, solutions can be obtained with a computer, but explicit rate equations provide more insight into the reactions. One therefore looks for approximate solutions of the equations. One of these is provided by the steady-state treatment, which is applicable if (and only if) the intermediates are species that can be present only at low concentrations. If this condition is satisfied by an intermediate, the rate of change of its concentration during the course of reaction is always small and, as a good approximation, can be assumed to be zero, which means that the intermediate exists in a steady state. This approximation may safely be applied to atoms and free radicals present as reaction intermediates. With this approximation it is usually possible to obtain a reliable approximate equation for the overall reaction rate in terms of reactant concentrations. If this agrees with the experimental behaviour, the mechanism is accepted.

One situation to which the steady-state treatment does not apply is when a reaction is an explosion. Explosions occur because the concentration of intermediates does not remain steady during the course of reaction but rises to a high value so that the reaction goes out of control. This occurs if the reaction mechanism involves a special kind of chain called a branching chain. In the hydrogen-oxygen explosion, for example, the following reaction is known to occur:

$$H + O_2 \rightarrow OH + O.$$

In this step a single chain carrier hydrogen atom has produced two chain carriers: a hydroxyl group and an oxygen atom. The number of chain carriers increases rapidly and leads to an explosion.

Theories of Reaction Rates

Two different theoretical approaches to chemical kinetics have led to an understanding of the details of how elementary chemical reactions occur. Both of these are based on the idea of potential-energy surfaces, which are models showing how the potential energy of a reaction system varies with certain critical interatomic distances. The course of an elementary reaction is represented by the movement of the system over the potential-energy surface. One theoretical approach to the problem involves studying the region of the potential-energy surface that corresponds to the highest point on the energy barrier that separates the reactants from the products. This approach is relatively simple and leads to explicit general expressions for the reaction rate. The second approach involves considering the dynamics of the motion of the system over the potential-energy surface.

Transition-state Theory

The idea of a potential-energy surface sprang from the ideas of Dutch physical chemist Jacobus Henricus van 't Hoff and Swedish physicist Svante August Arrhenius that were put forward to explain the effect of temperature on reaction rates. An important advance was made in 1931 by American chemist Henry Eyring and British chemist Michael Polanyi, who constructed, on the basis of quantum mechanics, a potential-energy surface for the simple reaction:

$$H^{\alpha} + H^{\beta}-H^{\gamma} \rightarrow H^{\alpha}-H^{\beta}-H^{\gamma} \rightarrow H^{\alpha}-H^{\beta} + H^{\gamma}.$$

For convenience the labels α, β, and γ are added as superscripts. When this reaction occurs, an atom H^{α} attacks a hydrogen molecule $H^{\beta}-H^{\gamma}$ and abstracts one of the hydrogen atoms from it. As the bond begins to form, the $H^{\beta}-H^{\gamma}$ bond becomes more and more extended and finally breaks. Somewhere along the reaction path, there is a particular intermediate state corresponding to the maximum value of the potential energy.

This particular intermediate state is usually designated by the superscript ‡. It is known as an activated complex and plays an important role in what has come to be called transition-state theory, developed independently in 1935 by Eyring, Polanyi, and English physical chemist M.G. Evans. The essential feature of the theory is that the activated complexes are considered to be formed from the reactants in a state in which they are in equilibrium with the reactants. Thus, the above reaction can be written as:

$$H^{\alpha} + H^{\beta}-H^{\gamma} \rightarrow H^{\alpha}-H^{\beta}-H^{\gamma\ddagger} \text{ (at equilibrium with } H^{\alpha} + H^{\beta}-H^{\gamma}) \rightarrow H^{\alpha}-H^{\beta} + H^{\gamma}.$$

Since the activated complexes are in equilibrium, their concentration can be expressed in terms of the concentrations of the reactants. The reaction rate is this concentration multiplied by the frequency with which they form products, which is known from kinetic theory. Despite the approximations involved in transition-state theory, it has been successful in providing an insight into how chemical reactions occur and how their rates depend on various factors.

Molecular Dynamics

The second theoretical approach to chemical kinetics is referred to as molecular dynamics, or reaction dynamics. It is a more detailed treatment of reactions and is designed to investigate the atomic motions that occur during a chemical reaction and the quantum states of the reactant and product molecules. Such studies are important in testing the validity of transition-state theory and similar treatments. Also, there are important practical applications of kinetics, such as reactions occurring in lasers, for which information about the energy states of the products of a chemical reaction is needed; this information is not provided by transition-state theory but is an important outcome of molecular dynamics.

Consider a simple reaction of the type A + B−C → Y + Z, where A is an atom and B−C is a diatomic molecule. A dynamical calculation would first involve calculating, using quantum mechanics, a potential-energy surface that gives the potential energy corresponding to a set of initial configurations. One can make dynamical calculations for a variety of vibrational states of the reactant B−C and for a variety of translational energies. Ideally the calculations would be based on quantum mechanics, but this proves difficult, and often classical mechanics is used. Except for certain types of reactions where quantum effects are important, it appears that for many reaction systems the error involved in neglecting quantum effects is not great. In principle the dynamical calculations should give more-reliable results than any other treatment; however, for any but the simplest reactions, the computer calculations are time-consuming, and, since many approximations must be made to save computer time, there is often some uncertainty about the results.

Much work has been done along these lines, and on the whole the agreement with both experiment and transition-state theory is satisfactory. In addition, some calculations suggest important generalizations about reactions. For example, the form of the potential-energy surface greatly influences whether energy released in a reaction resides in the vibrations of the product molecules or in their kinetic energy of translation. Experimental studies of simple reactions, particularly some that involve the study of light emission in reactions (chemiluminescence) and that use narrow molecular beams, have also contributed to knowledge of chemical reactions.

Rate Law

The relationship between the rate of a reaction and the concentrations of reactants is expressed by a rate law. For example, the rate of the gas-phase decomposition of dinitrogen pentoxide,

$$2N_2O_5 \rightleftharpoons 4NO_2 + O_2$$

is directly proportional to the concentration of N_2O_5:

$$\text{rate} = k[N_2O_5]$$

Care must be taken not to confuse equilibrium constant expressions with rate law expressions. The expression for K_{eq} can always be written by inspecting the balanced reaction equation and contains a term for each species of the reaction (raised to the power of its coefficient) whose concentration changes during the reaction. The equilibrium constant for the above reaction is given below:

$$K_{eq} = \frac{[NO_2]^4[O_2]}{[N_2O_5]^2}$$

In contrast, the expression for the rate law generally bears no relation to the reaction equation, and must be determined experimentally.

More generally, for a reaction of the form

$$n_A A + n_B B + \ldots \rightarrow \text{products}$$

with no intermediate steps, the rate law is given by the following:

$$\text{rate} = k[A]^a[B]^b$$

in which the exponents a and b are usually (but not always) integers and, it must be emphasized, bear no relation to the coefficients n_A and n_B. Because the rate of a reaction has dimensions of concentration per unit time, the dimensions of the rate constant k depend on the exponents of the concentration terms in the rate law. If p is the sum of the exponents of the concentration terms in the rate law,

$$p = a + b + \ldots$$

then k has dimensions of $\frac{\text{concentration}^{1\text{-p}}}{\text{time}}$.

How quickly a reaction proceeds depends on the reaction mechanism—the step-by-step molecular pathway leading from reactants to products. Chemical kinetics is concerned with how chemical reaction rates are measured, how they can be predicted, and how reaction rate data is used to deduce probable reactions.

Consider the following reaction:

$$2A + B \rightarrow C$$

The reaction rate is expressed as a derivative of the concentration of reactant A or product C, with respect to time, t:

$$\text{rate} = \lim_{\Delta t \to 0} -\frac{\Delta[A]}{\Delta t}$$

$$= -\frac{d[A]}{dt}$$

$$\text{rate} = \lim_{\Delta t \to 0} \frac{\Delta[C]}{\Delta t}$$

$$= \frac{d[C]}{dt}$$

In this reaction, one mole of C is produced from every 2 moles of A and one mole of B. The rate of this reaction may be described in terms of either the disappearance of reactants over time, or the appearance of products over time:

$$\text{rate} = \frac{\text{decrease in concentration of reactants}}{\text{time}}$$

$$= \frac{\text{increase in concentration of reactants}}{\text{time}}$$

Because the concentration of a reactant decreases during the reaction, a negative sign is placed before a rate that is expressed in terms of reactants to make the rate positive. For the reaction above, the rate of reaction with respect to A is $-\Delta[A]/\Delta t$, the rate with respect to B is $-\Delta[B]/\Delta t$, and the rate with respect to C is $\Delta[C]/\Delta t$. In this particular reaction, the three rates are not equal. According to the stoichiometry of the reaction, A is used up twice as fast as B, and A is consumed twice as fast as C is produced. To show a standard rate of reaction in which the rates with respect to all substances are equal, the rate for each substance should be divided by its stoichiometric coefficient:

$$\text{rate} = -\frac{1}{2}\frac{\Delta[A]}{\Delta t} = -\frac{\Delta[B]}{\Delta t} = \frac{\Delta[C]}{\Delta t}$$

Differential and Integral Rate Laws

Measuring instantaneous rates is the most direct way of determining the rate law of a reaction, but it is not always convenient, and it may not be possible to do so with precision.

- If the reaction is very fast, its rate may change more rapidly than the time required to measure it; the reaction may be finished before even an initial rate can be observed.
- In the case of very slow reactions, observable changes in concentrations occur so slowly that the observation of a truly "instantaneous" rate becomes impractical.

The ordinary rate law (more precisely known as the instantaneous or *differential* rate law) shows how the rate of a reaction depends on the concentrations of the reactants. However, for many practical purposes, it is more important to know how the concentrations of reactants (and products) change with time. For example, when carrying out a reaction on an industrial scale, it is important to know how long it will take for, as an example, 95% of the reactants to be converted into products.

For nearly all forward, irreversible reactions, the rate is proportional to the product of the concentrations of only the reactants, each raised to an exponent. For the general reaction:

$$aA + bB \rightarrow cC + dD$$

the rate is proportional to $[A]^m[B]^n$

$$\text{rate} = k[A]^m[B]^n$$

This expression is the rate law for the general reaction above, where k is the rate constant. Multiplying the units of k by the concentration factors raised to the appropriate powers give the rate in units of concentration/time.

The dependence of the rate of reaction on the reactant concentrations can often be expressed as a direct proportionality, in which the concentrations may be raised to be the zeroth, first, or second power. The exponent is known as the order of the reaction with respect to that substance. In the reaction above, the overall order of reaction is given by the following:

$$\text{order} = m + n$$

The order of the chemical equation can only be determined experimentally, i.e., mm and n cannot be determined from a balanced chemical equation alone (e.g., equation $aA + bB \rightarrow cC + dD$). The overall order of a reaction is the sum of the orders with respect to the sum of the exponents (equation order = $m + n$). Furthermore, the order of a reaction is stated with respect to a named substance in the reaction. The exponents in the rate law are not equal to the stoichiometric coefficients unless the reaction actually occurs via a single step mechanism (an elementary step); however, the exponents are equal to the stoichiometric coefficients of the rate-determining step. In general, the rate law can calculate the rate of reaction from known concentrations for reactants and derive an equation that expresses a reactant as a function of time.

The proportionality factor k, called the rate constant, is a constant at a fixed temperature; nonetheless, the rate constant varies with temperature. There are dimensions to k and that be determined with simple dimensional analysis of the particular rate law. The units should be expressed when the k values are tabulated. The higher the k value, the faster the reaction proceeds.

Experimental Determination of Rate Law

The values of k, mm, and n in the rate law equation must be determined experimentally for a given reaction at a given temperature. The rate is usually measured as a function of the initial concentrations of the reactants, A and B.

Given the data below, find the rate law for the following reaction at 300 K and the corresponding rate constant.

$$A+B \rightarrow C+D$$

Trial	$[A]_{initial}$ (M)	$[B]_{initial}$ (M)	$r_{initial}$ (M/sec)
1	1	1	2
2	1	2	8.1
3	2	2	15.9

Solution:

First, look for two trials in which the concentrations of all but one of the substances are held constant.

In trials 1 and 2, the concentration of A is kept constant while the concentration of B is doubled. The rate increases by a factor of approximately 4. Write down the rate expression of the two trials.

- Trial 1: $r_1 = k[A]^x[B]^y = k(1.00)^x(1.00)^y$
- Trial 2: $r_2 = k[A]^x[B]^y = k(1.00)^x(2.00)^y$

Divide the second equation by the first, which yields:

$$4 = (2.00)^y$$

$$y=2$$

In trials 2 and 3, the concentration of B is kept constant while the concentration of A is doubled; the rate is increased by a factor of approximately 2. The rate expressions of the two trials are as follows:

- Trial 2: $r_2 = k[A]^x[B]^y = k(1.00)^x(2.00)^y$
- Trial 3: $r_3 = k[A]^x[B]^y = k(2.00)^x(1.00)^y$

Divide the second equation by the third which yields:

$$2=(2.00)^x$$

$$x=1$$

Therefore, the rate law is:

$$r=2.0[A][B]^2$$

The order of the reaction with respect to A is 1 and with respect to B is 2; the overall reaction order is as follows:

$$1 + 2 = 3$$

To calculate k, substitute the values from any one of the above trials into the rate law:

$$2.0\ M/sec = k\ (1.00\ M)(1.00M)^2$$

$$k = 2.0M^{-2}\ sec^{-1}$$

Order of Reactions

Chemical reactions are often classified as zero-order, first-order, second-order, mixed-order, or higher-order. The general reaction aA + bB → cC + dD will be used in the following discussion. First, the meanings of these orders are defined in terms of initial rate of reaction effect:

- Zero-order in the reactant: There is no effect on the initial rate of reaction.
- First-order in the reactant: The initial rate of reaction doubles when the reactant is doubled.
- Second order in the reactant: The initial rate of the reaction quadruples when the reactant is doubled.
- Third order in the reactant: The initial rate of reaction increases eightfold when the reactant is doubled.

Zero-order Reactions

A zero-order reaction has a constant rate that is independent of the reactant's concentrations. The rate law is as follows:

$$\text{rate} = k$$

where k has units of Ms^{-1}. In other words, a zero-order reaction has a rate law in which the sum of the exponents is equal to zero. A reaction is zero order if concentration data are plotted versus time and the result is a straight line. The slope of this resulting line is the negative of the zero order rate constant, –k.

At times, chemists and researchers are also concerned with the relationship between the concentration of a reactant and time. An expression that shows this relationship is called an integrated rate law, in which the equation expresses the concentration of a reactant as a function of time (remember, each order of reaction has its own unique integrated rate law). The integrated rate law of a zero-order reaction is given below:

$$[A]_t = -kt + [A]_0$$

Notice, however, that this model is not entirely accurate because this equation predicts negative concentrations at sufficiently large times. In other words, if one were to graph the concentration of AA as a function of time, at some point, the line would cross the x-axis. This is physically impossible because concentrations cannot be negative. Nevertheless, this model is a sufficient model for ranges of time where concentration is predicted to be greater than zero.

The half life ($t_{1/2}$) of a reaction is the time required for the concentration of the radioactive substance to decrease to one-half of its original value. The half-life of a zero-order reaction can be derived as follows.

For a reaction involving reactant A and from the definition of a half-life, $t_{1/2}$ is the time it takes for half of the initial concentration of reactant A to react. These new conditions can be substituted into the integrated rate law form to obtain the following:

$$\frac{1}{2}[A]_0 = -kt_{1/2} + [A]_0$$

Solving for $t_{1/2}$ gives the following:

$$t_{1/2} = \frac{[A]_0}{2k}$$

First-order Reactions

A first-order reaction has a rate proportional to the concentration of one reactant:

$$\text{rate} = k[A] \quad \text{or} \quad \text{rate} = k[B]$$

First-order rate constants have units of sec^{-1}. In other words, a first-order reaction has a rate law in which the sum of the exponents is equal to 1.

The integrated rate law of a first-order reaction is the following:

$$\ln[A]_t = -kt + \ln[A]_0$$

$$\ln\frac{[A]_t}{[A]_0} = -kt$$

$$[A] = [A]_0 e^{-kt}$$

Moreover, a first-order reaction can be determined by plotting a graph of ln[*A*] vs. time *t*: a straight line is produced with slope -k.

The classic example of a first-order reaction is the process of radioactive decay. The concentration of radioactive substance A at any time t can be expressed mathematically below:

$$[A]_t = [A]_0 e^{-kt}$$

where $[A]_0$ is initial concentration of A, $[A]_t$ is the concentration of A at time t, k is the rate constant, and t is the elapsed time.

Second-order Reactions

A second-order reaction has a rate proportional to the product of the concentrations of two reactants, or to the square of the concentration of a single reactant. For example, each of the equations below describe a second-order reaction:

$$\text{rate} = k[A]^2$$

rate = $k[B]^2$

rate = $k[A][B]$

In other words, a second-order reaction has a rate law in which the sum of the exponents is equal to 2.

The integrated rate law of a second-order reaction is as follow:

$$\frac{1}{[A]} = \frac{1}{[A]_0} + kt$$

The half-life of a second-order reaction is given below:

$$t_{1/2} = \frac{1}{k[A]_0}$$

Enzyme Kinetics

Enzyme kinetics is the study of the chemical reactions that are catalysed by enzymes. In enzyme kinetics, the reaction rate is measured and the effects of varying the conditions of the reaction are investigated. Studying an enzyme's kinetics in this way can reveal the catalytic mechanism of this enzyme, its role in metabolism, how its activity is controlled, and how a drug or an agonist might inhibit the enzyme.

Enzymes are usually protein molecules that manipulate other molecules—the enzymes' substrates. These target molecules bind to an enzyme's active site and are transformed into products through a series of steps known as the enzymatic mechanism.

$$E + S \rightleftarrows ES \rightleftarrows ES^* \rightleftarrows EP \rightleftarrows E + P$$

These mechanisms can be divided into single-substrate and multiple-substrate mechanisms. Kinetic studies on enzymes that only bind one substrate, such as triosephosphate isomerase, aim to measure the affinity with which the enzyme binds this substrate and the turnover rate. Some other examples of enzymes are phosphofructokinase and hexokinase, both of which are important for cellular respiration (glycolysis).

When enzymes bind multiple substrates, such as dihydrofolate reductase (shown right), enzyme kinetics can also show the sequence in which these substrates bind and the sequence in which products are released. An example of enzymes that bind a single substrate and release multiple products are proteases, which cleave one protein substrate into two polypeptide products. Others join two substrates together, such as DNA polymerase linking a nucleotide to DNA. Although these mechanisms are often a complex series of steps, there is typically one *rate-determining step* that determines the overall kinetics. This rate-determining step may be a chemical reaction or a conformational change of the enzyme or substrates, such as those involved in the release of product(s) from the enzyme.

Knowledge of the enzyme's structure is helpful in interpreting kinetic data. For example, the structure can suggest how substrates and products bind during catalysis; what changes occur during the reaction; and even the role of particular amino acid residues in the mechanism. Some enzymes change shape significantly during the mechanism; in such cases, it is helpful to determine the enzyme structure with and without bound substrate analogues that do not undergo the enzymatic reaction.

Not all biological catalysts are protein enzymes; RNA-based catalysts such as ribozymes and ribosomes are essential to many cellular functions, such as RNA splicing and translation. The main difference between ribozymes and enzymes is that RNA catalysts are composed of nucleotides, whereas enzymes are composed of amino acids. Ribozymes also perform a more limited set of reactions, although their reaction mechanisms and kinetics can be analysed and classified by the same methods.

Principles

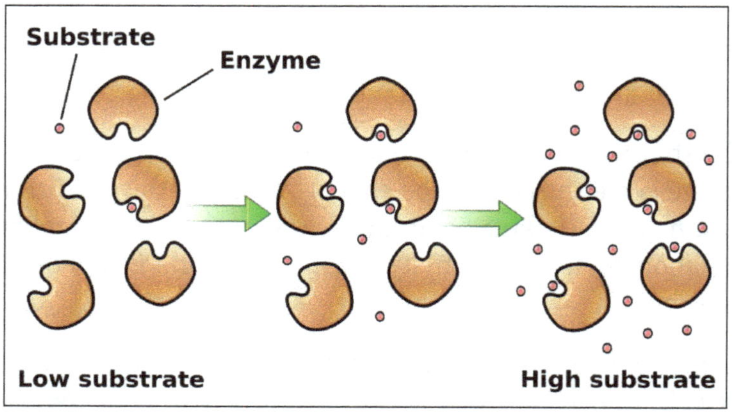

As larger amounts of substrate are added to a reaction, the available enzyme binding sites become filled to the limit of V_{max}. Beyond this limit the enzyme is saturated with substrate and the reaction rate ceases to increase.

The reaction catalysed by an enzyme uses exactly the same reactants and produces exactly the same products as the uncatalysed reaction. Like other catalysts, enzymes do not alter the position of equilibrium between substrates and products. However, unlike uncatalysed chemical reactions, enzyme-catalysed reactions display saturation kinetics. For a given enzyme concentration and for relatively low substrate concentrations, the reaction rate increases linearly with substrate concentration; the enzyme molecules are largely free to catalyse the reaction, and increasing substrate concentration means an increasing rate at which the enzyme and substrate molecules encounter one another. However, at relatively high substrate concentrations, the reaction rate asymptotically approaches the theoretical maximum; the enzyme active sites are almost all occupied by substrates resulting in saturation, and the reaction rate is determined by the intrinsic turnover rate of the enzyme. The substrate concentration midway between these two limiting cases is denoted by K_M. Thus, K_M is the substrate concentration at which the reaction velocity is half of the maximum velocity.

The two most important kinetic properties of an enzyme are how easily the enzyme becomes saturated with a particular substrate, and the maximum rate it can achieve. Knowing these properties suggests what an enzyme might do in the cell and can show how the enzyme will respond to changes in these conditions.

Enzyme Assays

Enzyme assays are laboratory procedures that measure the rate of enzyme reactions. Since enzymes are not consumed by the reactions they catalyse, enzyme assays usually follow changes in the concentration of either substrates or products to measure the rate of reaction. There are many methods of measurement. Spectrophotometric assays observe change in the absorbance of light between products and reactants; radiometric assays involve the incorporation or release of radioactivity to measure the amount of product made over time. Spectrophotometric assays are most convenient since they allow the rate of the reaction to be measured continuously. Although radiometric assays require the removal and counting of samples (i.e., they are discontinuous assays) they are usually extremely sensitive and can measure very low levels of enzyme activity. An analogous approach is to use mass spectrometry to monitor the incorporation or release of stable isotopes as substrate is converted into product.

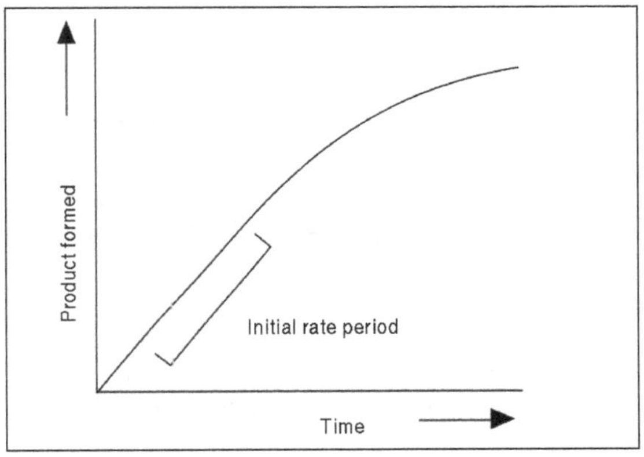

Progress curve for an enzyme reaction. The slope in the initial rate period is the initial rate of reaction v. The Michaelis–Menten equation describes how this slope varies with the concentration of substrate.

The most sensitive enzyme assays use lasers focused through a microscope to observe changes in single enzyme molecules as they catalyse their reactions. These measurements either use changes in the fluorescence of cofactors during an enzyme's reaction mechanism, or of fluorescent dyes added onto specific sites of the protein to report movements that occur during catalysis. These studies are providing a new view of the kinetics and dynamics of single enzymes, as opposed to traditional enzyme kinetics, which observes the average behaviour of populations of millions of enzyme molecules.

An example progress curve for an enzyme assay is shown above. The enzyme produces product at an initial rate that is approximately linear for a short period after the start of the reaction. As the reaction proceeds and substrate is consumed, the rate continuously slows (so long as substrate is not still at saturating levels). To measure the initial (and maximal) rate, enzyme assays are typically carried out while the reaction has progressed only a few percent towards total completion. The length of the initial rate period depends on the assay conditions and can range from milliseconds to hours. However, equipment for rapidly mixing liquids allows fast kinetic measurements on initial rates of less than one second. These very rapid assays are essential for measuring pre-steady-state kinetics.

Most enzyme kinetics studies concentrate on this initial, approximately linear part of enzyme reactions. However, it is also possible to measure the complete reaction curve and fit this data to a non-linear rate equation. This way of measuring enzyme reactions is called progress-curve analysis. This approach is useful as an alternative to rapid kinetics when the initial rate is too fast to measure accurately.

Single-substrate Reactions

Enzymes with single-substrate mechanisms include isomerases such as triosephosphateisomerase or bisphosphoglycerate mutase, intramolecular lyases such as adenylate cyclase and the hammerhead ribozyme, an RNA lyase. However, some enzymes that only have a single substrate do not fall into this category of mechanisms. Catalase is an example of this, as the enzyme reacts with a first molecule of hydrogen peroxide substrate, becomes oxidised and is then reduced by a second molecule of substrate. Although a single substrate is involved, the existence of a modified enzyme intermediate means that the mechanism of catalase is actually a ping–pong mechanism.

Michaelis–Menten Kinetics

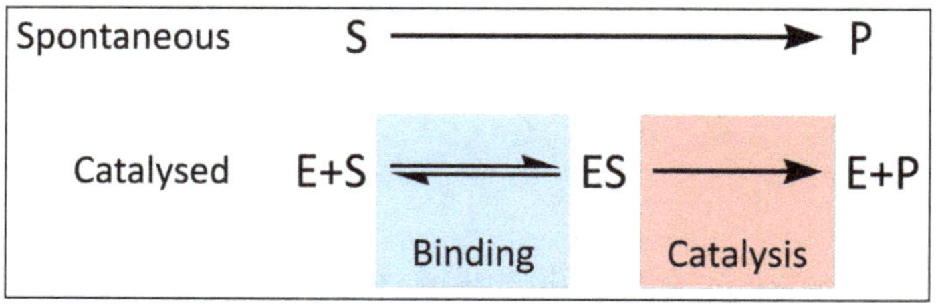

A chemical reaction mechanism with or without enzyme catalysis. The enzyme (E) binds substrate (S) to produce product (P).

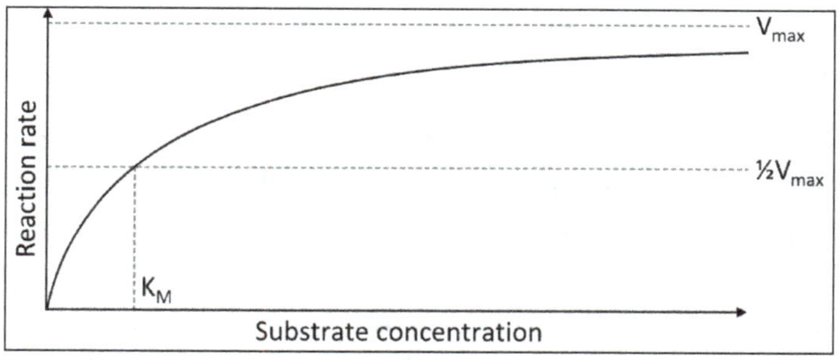

Saturation curve for an enzyme reaction showing the relation between the substrate concentration and reaction rate.

As enzyme-catalysed reactions are saturable, their rate of catalysis does not show a linear response to increasing substrate. If the initial rate of the reaction is measured over a range of substrate concentrations (denoted as [S]), the initial reaction rate (v_0) increases as [S] increases, as shown on the right. However, as [S] gets higher, the enzyme becomes saturated with substrate and the initial rate reaches V_{max}, the enzyme's maximum rate.

The Michaelis–Menten kinetic model of a single-substrate reaction is shown on the right. There is an initial bimolecular reaction between the enzyme E and substrate S to form the enzyme–substrate complex ES. The rate of enzymatic reaction increases with the increase of the substrate concentration up to a certain level called V_{max}; at V_{max}, increase in substrate concentration does not cause any increase in reaction rate as there is no more enzyme (E) available for reacting with substrate (S). Here, the rate of reaction becomes dependent on the ES complex and the reaction becomes a unimolecular reaction with an order of zero. Though the enzymatic mechanism for the unimolecular reaction $ES \xrightarrow{k_{cat}} E + P$ can be quite complex, there is typically one rate-determining enzymatic step that allows this reaction to be modelled as a single catalytic step with an apparent unimolecular rate constant k_{cat}. If the reaction path proceeds over one or several intermediates, k_{cat} will be a function of several elementary rate constants, whereas in the simplest case of a single elementary reaction (e.g. no intermediates) it will be identical to the elementary unimolecular rate constant k_2. The apparent unimolecular rate constant k_{cat} is also called *turnover number* and denotes the maximum number of enzymatic reactions catalysed per second.

The Michaelis–Menten equation describes how the (initial) reaction rate v_0 depends on the position of the substrate-binding equilibrium and the rate constant k_2:

$$v_0 = \frac{V_{\max}[S]}{K_M + [S]} \qquad \text{(Michaelis–Menten equation)}$$

with the constants,

$$K_M \stackrel{\text{def}}{=} \frac{k_2 + k_{-1}}{k_1} \approx K_D$$

$$V_{\max} \stackrel{\text{def}}{=} k_{cat}[E]_{tot}$$

This Michaelis–Menten equation is the basis for most single-substrate enzyme kinetics. Two crucial assumptions underlie this equation (apart from the general assumption about the mechanism only involving no intermediate or product inhibition, and there is no allostericity or cooperativity). The first assumption is the so-called quasi-steady-state assumption (or pseudo-steady-state hypothesis), namely that the concentration of the substrate-bound enzyme (and hence also the unbound enzyme) changes much more slowly than those of the product and substrate and thus the change over time of the complex can be set to zero $d[ES]/dt \stackrel{!}{=} 0$. The second assumption is that the total enzyme concentration does not change over time, thus $[E]_{tot} = [E] + [ES] \stackrel{!}{=} const.$

The Michaelis constant K_M is experimentally defined as the concentration at which the rate of the enzyme reaction is half V_{max}, which can be verified by substituting [S] = K_M into the Michaelis–Menten equation and can also be seen graphically. If the rate-determining enzymatic step is slow compared to substrate dissociation ($k_2 \ll k_{-1}$), the Michaelis constant K_M is roughly the dissociation constant K_D of the ES complex.

If $[S]$ is small compared to K_M then the term $[S]/(K_M + [S]) \approx [S]/K_M$ and also very little ES complex is formed, thus $[E]_{tot} \approx [E]$. Therefore, the rate of product formation is:

$$v_0 \approx \frac{k_{cat}}{K_M}[E][S] \qquad \textit{if } [S] \ll K_M$$

Thus the product formation rate depends on the enzyme concentration as well as on the substrate concentration, the equation resembles a bimolecular reaction with a corresponding pseudo-second order rate constant k_2 / K_M. This constant is a measure of catalytic efficiency. The most efficient enzymes reach a k_2 / K_M in the range of 10^8 – 10^{10} M^{-1} s^{-1}. These enzymes are so efficient they effectively catalyse a reaction each time they encounter a substrate molecule and have thus reached an upper theoretical limit for efficiency (diffusion limit); and are sometimes referred to as kinetically perfect enzymes. But most enzymes are far from perfect: the average values of k_2 / K_M and k_2 are about $10^5 s^{-1} M^{-1}$ and $10 s^{-1}$ respectively.

Direct use of the Michaelis–Menten Equation

The observed velocities predicted by the Michaelis–Menten equation can be used to directly model the time course disappearance of substrate and the production of product through incorporation of the Michaelis–Menten equation into the equation for first order chemical kinetics. This can only be achieved however if one recognises the problem associated with the use of Euler's number in the description of first order chemical kinetics. i.e. e^{-k} is a split constant that introduces a systematic error into calculations and can be rewritten as a single constant which represents the remaining substrate after each time period.

$$[S] = [S]_0 (1-k)^t$$

$$[S] = [S]_0 (1 - v / [S]_0)^t$$

$$[S] = [S]_0 (1 - (V_{\max} [S]_0 / (K_M + [S]_0) / [S]_0))^t$$

In 1983 Stuart Beal derived a closed form solution for the time course kinetics analysis of the Michaelis-Menten mechanism. The solution, known as the Schnell-Mendoza equation, has the form:

$$\frac{[S]}{K_M} = W\left[F(t)\right]$$

where W[ ] is the Lambert-W function. and where F(t) is:

$$F(t) = \frac{[S]_0}{K_M} \exp\left(\frac{[S]_0}{K_M} - \frac{V_{\max}}{K_M} t\right)$$

This equation is encompassed by the equation below, obtained by Berberan-Santos, which is also valid when the initial substrate concentration is close to that of enzyme,

$$\frac{[S]}{K_M} = W\left[F(t)\right] - \frac{V_{\max}}{k_{cat} K_M} \frac{W\left[F(t)\right]}{1 + W\left[F(t)\right]}$$

where W[ ] is again the Lambert-W function.

Linear Plots of the Michaelis–Menten Equation

The plot of v versus [S] above is not linear; although initially linear at low [S], it bends over to saturate at high [S]. Before the modern era of nonlinear curve-fitting on computers, this nonlinearity could make it difficult to estimate K_M and V_{max} accurately. Therefore, several researchers developed linearisations of the Michaelis–Menten equation, such as the Lineweaver–Burk plot, the Eadie–Hofstee diagram and the Hanes–Woolf plot. All of these linear representations can be useful for visualising data, but none should be used to determine kinetic parameters, as computer software is readily available that allows for more accurate determination by nonlinear regression methods.

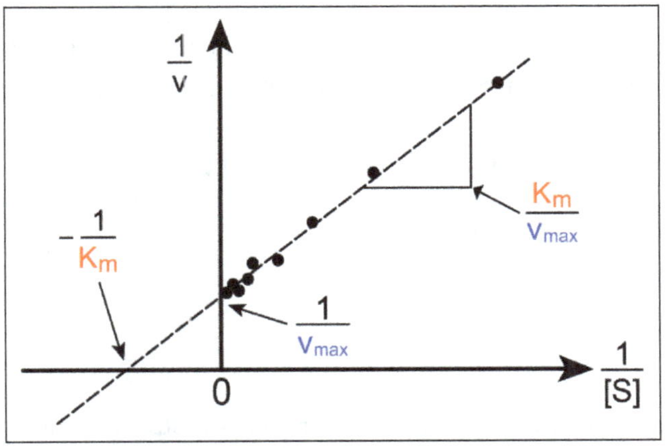

Lineweaver–Burk or double-reciprocal plot of kinetic data, showing the significance of the axis intercepts and gradient.

The Lineweaver–Burk plot or double reciprocal plot is a common way of illustrating kinetic data. This is produced by taking the reciprocal of both sides of the Michaelis–Menten equation. As shown on the right, this is a linear form of the Michaelis–Menten equation and produces a straight line with the equation y = mx + c with a y-intercept equivalent to $1/V_{max}$ and an x-intercept of the graph representing $-1/K_M$.

$$\frac{1}{v} = \frac{K_M}{V_{max}[S]} + \frac{1}{V_{max}}$$

Naturally, no experimental values can be taken at negative 1/[S]; the lower limiting value 1/[S] = 0 (the y-intercept) corresponds to an infinite substrate concentration, where $1/v=1/V_{max}$ as shown at the right; thus, the x-intercept is an extrapolation of the experimental data taken at positive concentrations. More generally, the Lineweaver–Burk plot skews the importance of measurements taken at low substrate concentrations and, thus, can yield inaccurate estimates of V_{max} and K_M. A more accurate linear plotting method is the Eadie–Hofstee plot. In this case, v is plotted against v/[S]. In the third common linear representation, the Hanes–Woolf plot, [S]/v is plotted against [S]. In general, data normalisation can help diminish the amount of experimental work and can increase the reliability of the output, and is suitable for both graphical and numerical analysis.

Practical Significance of Kinetic Constants

The study of enzyme kinetics is important for two basic reasons. Firstly, it helps explain how enzymes work, and secondly, it helps predict how enzymes behave in living organisms. The kinetic constants defined above, K_M and V_{max}, are critical to attempts to understand how enzymes work together to control metabolism.

Making these predictions is not trivial, even for simple systems. For example, oxaloacetate is formed by malate dehydrogenase within the mitochondrion. Oxaloacetate can then be consumed by citrate synthase, phosphoenolpyruvate carboxykinase or aspartate aminotransferase, feeding into the citric acid cycle, gluconeogenesis or aspartic acid biosynthesis, respectively. Being able to predict how much oxaloacetate goes into which pathway requires knowledge of the concentration of oxaloacetate as well as the concentration and kinetics of each of these enzymes. This aim of predicting the behaviour of metabolic pathways reaches its most complex expression in the synthesis of huge amounts of kinetic and gene expression data into mathematical models of entire organisms. Alternatively, one useful simplification of the metabolic modelling problem is to ignore the underlying enzyme kinetics and only rely on information about the reaction network's stoichiometry, a technique called flux balance analysis.

Michaelis–Menten Kinetics with Intermediate

One could also consider the less simple case:

$$E + S \underset{k_{-1}}{\overset{k_1}{\rightleftharpoons}} ES \overset{k_2}{\rightarrow} EI \overset{k_3}{\rightarrow} E + P$$

where a complex with the enzyme and an intermediate exists and the intermediate is converted into product in a second step. In this case we have a very similar equation:

$$v_0 = k_{cat} \frac{[S][E]_0}{K_M^{'} + [S]}$$

but the constants are different,

$$K_M^{'} \overset{\text{def}}{=} \frac{k_3}{k_2 + k_3} K_M = \frac{k_3}{k_2 + k_3} . \frac{k_2 + k_{-1}}{k_1}$$

$$k_{cat} \overset{\text{def}}{=} \frac{k_3 k_2}{k_2 + k_3}$$

We see that for the limiting case $k_3 \gg k_2$, thus when the last step from $\mathrm{EI} \rightarrow \mathrm{E+P}$ is much faster than the previous step, we get again the original equation. Mathematically we have then $K_M^{'} \approx K_M$ and $k_{cat} \approx k_2$.

Multi-substrate Reactions

Multi-substrate reactions follow complex rate equations that describe how the substrates bind and in what sequence. The analysis of these reactions is much simpler if the concentration of substrate A is kept constant and substrate B varied. Under these conditions, the enzyme behaves just like a single-substrate enzyme and a plot of v by [S] gives apparent K_M and V_{max} constants for substrate B. If a set of these measurements is performed at different fixed concentrations of A, these data can be used to work out what the mechanism of the reaction is. For an enzyme that takes two substrates A and B and turns them into two products P and Q, there are two types of mechanism: ternary complex and ping–pong.

Ternary-complex Mechanisms

In these enzymes, both substrates bind to the enzyme at the same time to produce an EAB ternary complex. The order of binding can either be random (in a random mechanism) or substrates have to bind in a particular sequence (in an ordered mechanism). When a set of v by [S] curves (fixed A, varying B) from an enzyme with a ternary-complex mechanism are plotted in a Lineweaver–Burk plot, the set of lines produced will intersect.

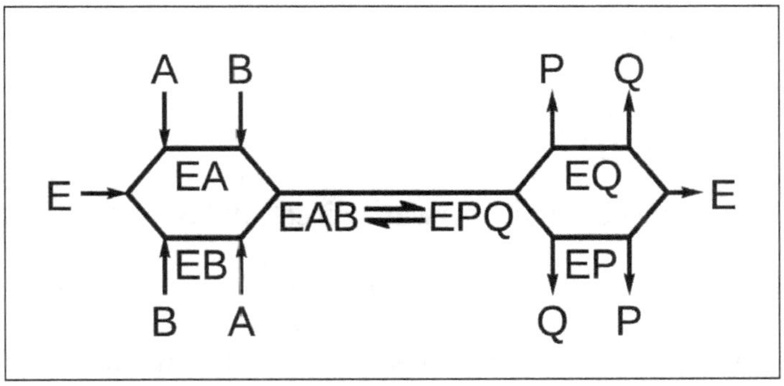

Random-order ternary-complex mechanism for an enzyme reaction. The reaction path is shown as a line and enzyme intermediates containing substrates A and B or products P and Q are written below the line.

Enzymes with ternary-complex mechanisms include glutathione *S*-transferase, dihydrofolate reductase and DNA polymerase.

Ping–pong Mechanisms

$$E \xrightarrow{\overset{A}{\downarrow}} EA \rightleftharpoons E^*P \xrightarrow{\overset{P}{\uparrow}} E^* \xrightarrow{\overset{B}{\downarrow}} E^*B \rightleftharpoons EQ \xrightarrow{\overset{Q}{\uparrow}} E$$

Ping–pong mechanism for an enzyme reaction. Intermediates contain substrates A and B or products P and Q.

As shown on the right, enzymes with a ping-pong mechanism can exist in two states, E and a chemically modified form of the enzyme E*; this modified enzyme is known as an intermediate. In such mechanisms, substrate A binds, changes the enzyme to E* by, for example, transferring a chemical group to the active site, and is then released. Only after the first substrate is released can substrate B bind and react with the modified enzyme, regenerating the unmodified E form. When a set of v by [S] curves (fixed A, varying B) from an enzyme with a ping–pong mechanism are plotted in a Lineweaver–Burk plot, a set of parallel lines will be produced. This is called a secondary plot.

Enzymes with ping–pong mechanisms include some oxidoreductases such as thioredoxin peroxidase, transferases such as acylneuraminate cytidylyltransferase and serine proteases such as trypsin and chymotrypsin. Serine proteases are a very common and diverse family of enzymes, including digestive enzymes (trypsin, chymotrypsin, and elastase), several enzymes of the blood clotting cascade and many others. In these serine proteases, the E* intermediate is an acyl-enzyme species formed by the attack of an active site serine residue on a peptide bond in a protein substrate.

Reversible Catalysis and Haldane Equation

We consider the case of an enzyme that can catalyse the reaction in both directions (not all enzymes can):

$$E + S \underset{k_{-1}}{\overset{k_1}{\rightleftharpoons}} ES \underset{k_{-2}}{\overset{k_{-2}}{\rightleftharpoons}} E + P$$

The steady-state, initial rate of the reaction is:

$$v_0 = \frac{d[\mathrm{P}]}{dt} = \frac{(k_1 k_2[\mathrm{S}] - k_{-1}k_{-2}[\mathrm{P}])[\mathrm{E}]_0}{k_{-1} + k_2 + k_1[\mathrm{S}] + k_{-2}[\mathrm{P}]}$$

v_0 is positive if the reaction proceed in the forward direction ($S \to P$) and negative otherwise. Equilibrium requires that $v = 0$,, which occurs when $\frac{[\mathrm{P}]_{\mathrm{eq}}}{[\mathrm{S}]_{\mathrm{eq}}} = \frac{k_1 k_2}{k_{-1}k_{-2}} = K_{\mathrm{eq}}$. This shows that thermodynamics forces a relation between the values of the 4 rate constants.

The values of the forward and backward maximal rates, obtained for $[\mathrm{S}] \to \infty$, $[\mathrm{P}] = 0$, and $[\mathrm{S}] = 0$, $[\mathrm{P}] \to \infty$, respectively, are $V^f_{\max} = k_2[\mathrm{E}]_{tot}$ and $V^b_{\max} = -k_{-1}[\mathrm{E}]_{tot}$, respectively. Their ratio is not equal to the equilibrium constant, which implies that thermodynamics does not constrain the ratio of the maximal rates. This explains that enzymes can be much "better catalysts" (in terms of maximal rates) in one particular direction of the reaction.

On can also derive the two Michaelis constants $K^S_M = (k_{-1} + k_2)/k_1$ and $K^P_M = (k_{-1} + k_2)/k_{-2}$. The Haldane equation is the relation $K_{\mathrm{eq}} = \frac{ü_{\mathrm{eq}}}{[\mathrm{S}]_{\mathrm{eq}}} = \frac{V^f_{ü} / K^S_M}{V^b_{\max} / K^P_M}$.

Therefore, thermodynamics constrains the ratio between the forward and backward $V_{\max}/K_M$ values, not the ratio of $V_{\max}$ values.

Non-Michaelis–Menten Kinetics

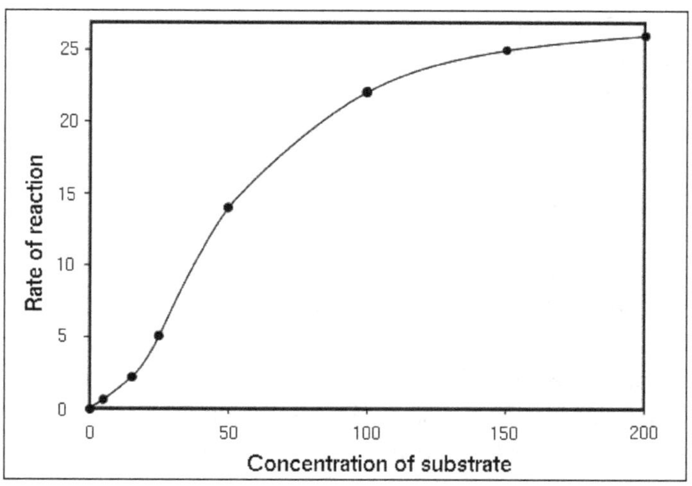

Saturation curve for an enzyme reaction showing sigmoid kinetics.

Some enzymes produce a sigmoid v by [S] plot, which often indicates cooperative binding of substrate to the active site. This means that the binding of one substrate molecule affects the binding of subsequent substrate molecules. This behavior is most common in multimeric enzymes with several interacting active sites. Here, the mechanism of cooperation is similar to that of hemoglobin, with binding of substrate to one active site altering the affinity of the other active sites for substrate molecules. Positive cooperativity occurs when binding of the first substrate molecule *increases* the affinity of the other active sites for substrate. Negative cooperativity occurs when binding of the first substrate *decreases* the affinity of the enzyme for other substrate molecules.

Allosteric enzymes include mammalian tyrosyl tRNA-synthetase, which shows negative cooperativity, and bacterial aspartate transcarbamoylase and phosphofructokinase, which show positive cooperativity.

Cooperativity is surprisingly common and can help regulate the responses of enzymes to changes in the concentrations of their substrates. Positive cooperativity makes enzymes much more sensitive to [S] and their activities can show large changes over a narrow range of substrate concentration. Conversely, negative cooperativity makes enzymes insensitive to small changes in [S].

The Hill equation (biochemistry) is often used to describe the degree of cooperativity quantitatively in non-Michaelis–Menten kinetics. The derived Hill coefficient n measures how much the binding of substrate to one active site affects the binding of substrate to the other active sites. A Hill coefficient of <1 indicates negative cooperativity and a coefficient of >1 indicates positive cooperativity.

Pre-steady-state Kinetics

In the first moment after an enzyme is mixed with substrate, no product has been formed and no intermediates exist. The study of the next few milliseconds of the reaction is called pre-steady-state kinetics. Pre-steady-state kinetics is therefore concerned with the formation and consumption of enzyme–substrate intermediates (such as ES or E*) until their steady-state concentrations are reached.

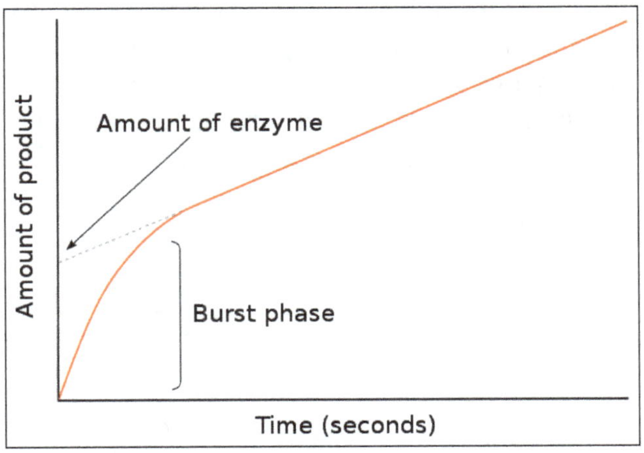

Pre-steady state progress curve, showing the burst phase of an enzyme reaction.

This approach was first applied to the hydrolysis reaction catalysed by chymotrypsin. Often, the detection of an intermediate is a vital piece of evidence in investigations of what mechanism an enzyme follows. For example, in the ping–pong mechanisms that are shown above, rapid kinetic measurements can follow the release of product P and measure the formation of the modified enzyme intermediate E*. In the case of chymotrypsin, this intermediate is formed by an attack on the substrate by the nucleophilic serine in the active site and the formation of the acyl-enzyme intermediate.

In the figure above, the enzyme produces E* rapidly in the first few seconds of the reaction. The rate then slows as steady state is reached. This rapid burst phase of the reaction measures a single turnover of the enzyme. Consequently, the amount of product released in this burst, shown as the intercept on the *y*-axis of the graph, also gives the amount of functional enzyme which is present in the assay.

Chemical Mechanism

An important goal of measuring enzyme kinetics is to determine the chemical mechanism of an enzyme reaction, i.e., the sequence of chemical steps that transform substrate into product. The kinetic approaches discussed above will show at what rates intermediates are formed and inter-converted, but they cannot identify exactly what these intermediates are.

Kinetic measurements taken under various solution conditions or on slightly modified enzymes or substrates often shed light on this chemical mechanism, as they reveal the rate-determining step or intermediates in the reaction. For example, the breaking of a covalent bond to a hydrogen atom is a common rate-determining step. Which of the possible hydrogen transfers is rate determining can be shown by measuring the kinetic effects of substituting each hydrogen by deuterium, its stable isotope. The rate will change when the critical hydrogen is replaced, due to a primary kinetic isotope effect, which occurs because bonds to deuterium are harder to break than bonds to hydrogen. It is also possible to measure similar effects with other isotope substitutions, such as $^{13}C/^{12}C$ and $^{18}O/^{16}O$, but these effects are more subtle.

Isotopes can also be used to reveal the fate of various parts of the substrate molecules in the final products. For example, it is sometimes difficult to discern the origin of an oxygen atom in the final product; since it may have come from water or from part of the substrate. This may be determined by systematically substituting oxygen's stable isotope ^{18}O into the various molecules that participate in the reaction and checking for the isotope in the product. The chemical mechanism can also be elucidated by examining the kinetics and isotope effects under different pH conditions, by altering the metal ions or other bound cofactors, by site-directed mutagenesis of conserved amino acid residues, or by studying the behaviour of the enzyme in the presence of analogues of the substrate.

Enzyme Inhibition and Activation

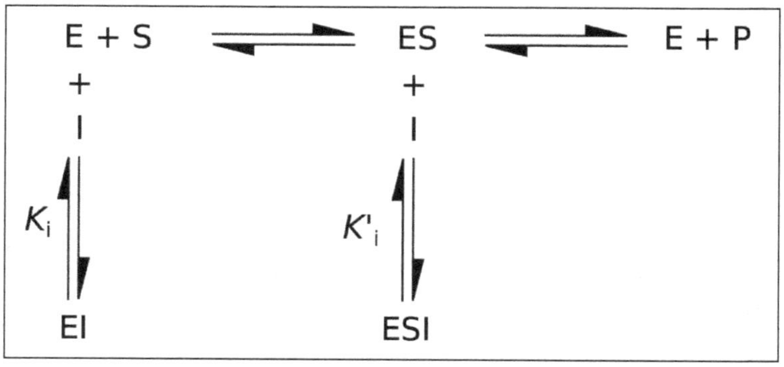

Kinetic scheme for reversible enzyme inhibitors.

Enzyme inhibitors are molecules that reduce or abolish enzyme activity, while enzyme activators are molecules that increase the catalytic rate of enzymes. These interactions can be either reversible (i.e., removal of the inhibitor restores enzyme activity) or irreversible (i.e., the inhibitor permanently inactivates the enzyme).

Reversible Inhibitors

Traditionally reversible enzyme inhibitors have been classified as competitive, uncompetitive, or non-competitive, according to their effects on K_M and V_{max}. These different effects result from the inhibitor binding to the enzyme E, to the enzyme–substrate complex ES, or to both, respectively. The division of these classes arises from a problem in their derivation and results in the need to use two different binding constants for one binding event. The binding of an inhibitor and its effect on the enzymatic activity are two distinctly different things, another problem the traditional equations fail to acknowledge. In noncompetitive inhibition the binding of the inhibitor results in 100% inhibition of the enzyme only, and fails to consider the possibility of anything in between. In noncompetitive inhibition, the inhibitor will bind to an enzyme at its allosteric site; therefore, the binding affinity, or inverse of K_M, of the substrate with the enzyme will remain the same. On the other hand, the Vmax will decrease relative to an uninhibited enzyme. On a Lineweaver-Burk plot, the presence of a noncompetitive inhibitor is illustrated by a change in the y-intercept, defined as 1/Vmax. The x-intercept, defined as $-1/K_M$, will remain the same. In competitive inhibition, the inhibitor will bind to an enzyme at the active site, competing with the substrate. As a result, the K_M will increase and the Vmax will remain the same. The common form of the inhibitory term also obscures the relationship between the inhibitor binding to the enzyme and its relationship to any other binding term be it the Michaelis–Menten equation or a dose response curve associated with ligand receptor binding. To demonstrate the relationship the following rearrangement can be made:

$$\frac{V_{max}}{1+\frac{[I]}{K_i}} = \frac{V_{max}}{\frac{[I]+K_i}{K_i}}$$

Adding zero to the bottom ([I]-[I]),

$$\frac{V_{max}}{\frac{[I]+K_i}{[I]+K_i-[I]}}$$

Dividing by $[I]+K_i$,

$$\frac{V_{max}}{\frac{1}{1-\frac{[I]}{[I]+K_i}}} = V_{max} - V_{max}\frac{[I]}{[I]+K_i}$$

This notation demonstrates that similar to the Michaelis–Menten equation, where the rate of reaction depends on the percent of the enzyme population interacting with substrate, the effect of the inhibitor is a result of the percent of the enzyme population interacting with inhibitor. The only problem with this equation in its present form is that it assumes absolute inhibition of the enzyme with inhibitor binding, when in fact there can be a wide range of effects anywhere from 100% inhibition of substrate turn over to just >0%. To account for this the equation can be easily modified to allow for different degrees of inhibition by including a delta V_{max} term:

$$V_{max} - \Delta V_{max} \frac{[I]}{[I]+K_i}$$

or

$$V_{max1} - (V_{max1} - V_{max2}) \frac{[I]}{[I]+K_i}$$

This term can then define the residual enzymatic activity present when the inhibitor is interacting with individual enzymes in the population. However the inclusion of this term has the added value of allowing for the possibility of activation if the secondary V_{max} term turns out to be higher than the initial term. To account for the possibly of activation as well the notation can then be rewritten replacing the inhibitor "I" with a modifier term denoted here as "X":

$$V_{max1} - (V_{max1} - V_{max2}) \frac{[X]}{[X]+K_x}$$

While this terminology results in a simplified way of dealing with kinetic effects relating to the maximum velocity of the Michaelis–Menten equation, it highlights potential problems with the term used to describe effects relating to the K_M. The K_M relating to the affinity of the enzyme for the substrate should in most cases relate to potential changes in the binding site of the enzyme which would directly result from enzyme inhibitor interactions. As such a term similar to the one proposed above to modulate V_{max} should be appropriate in most situations:

$$K_{m1} - (K_{m1} - K_{m2}) \frac{[X]}{[X]+K_x}$$

Irreversible Inhibitors

Enzyme inhibitors can also irreversibly inactivate enzymes, usually by covalently modifying active site residues. These reactions, which may be called suicide substrates, follow exponential decay functions and are usually saturable. Below saturation, they follow first order kinetics with respect to inhibitor. Irreversible inhibition could be classified into two distinct types. Affinity labelling is a type of irreversible inhibition where a functional group that is highly reactive modifies a catalytically critical residue on the protein of interest to bring about inhibition. Mechanism-based inhibition, on the other hand, involves binding of the inhibitor followed by enzyme mediated alterations that transform the latter into a reactive group that irreversibly modifies the enzyme.

Philosophical Discourse on Reversibility and Irreversibility of Inhibition

The concept of reversibility (or irreversibility) is a purely theoretical construct exclusively dependent on the time-frame of the assay, i.e., a reversible assay involving association and dissociation of the inhibitor molecule in the minute timescales would seem irreversible if an assay assess the outcome in the seconds and vice versa. There is a continuum of inhibitor behaviors spanning reversibility and irreversibility at a given non-arbitrary assay time frame. There are inhibitors that show slow-onset behavior and most of these inhibitors, invariably, also show tight-binding to the protein target of interest.

Mechanisms of Catalysis

The favoured model for the enzyme–substrate interaction is the induced fit model. This model proposes that the initial interaction between enzyme and substrate is relatively weak, but that these weak interactions rapidly induce conformational changes in the enzyme that strengthen binding. These conformational changes also bring catalytic residues in the active site close to the chemical bonds in the substrate that will be altered in the reaction. Conformational changes can be measured using circular dichroism or dual polarisation interferometry. After binding takes place, one or more mechanisms of catalysis lower the energy of the reaction's transition state by providing an alternative chemical pathway for the reaction. Mechanisms of catalysis include catalysis by bond strain; by proximity and orientation; by active-site proton donors or acceptors; covalent catalysis and quantum tunnelling.

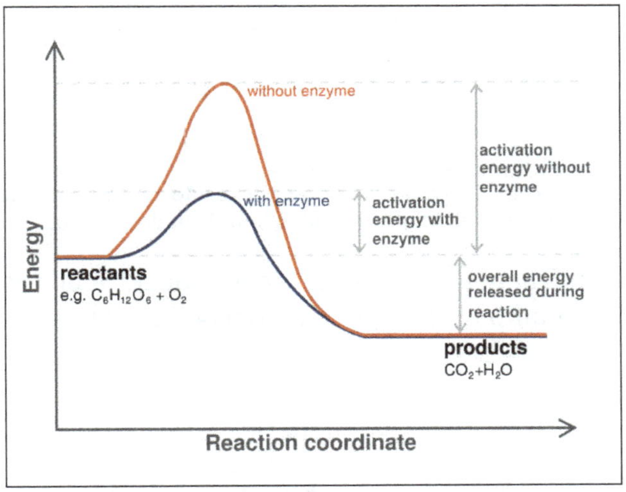

The energy variation as a function of reaction coordinate shows the stabilisation of the transition state by an enzyme.

Enzyme kinetics cannot prove which modes of catalysis are used by an enzyme. However, some kinetic data can suggest possibilities to be examined by other techniques. For example, a ping–pong mechanism with burst-phase pre-steady-state kinetics would suggest covalent catalysis might be important in this enzyme's mechanism. Alternatively, the observation of a strong pH effect on V_{max} but not K_M might indicate that a residue in the active site needs to be in a particular ionisation state for catalysis to occur.

Chemical Reactor

In chemical engineering, chemical reactors are vessels designed to contain chemical reactions. One example is a pressure reactor. The design of a chemical reactor deals with multiple aspects of chemical engineering. Chemical engineers design reactors to maximize net present value for the given reaction. Designers ensure that the reaction proceeds with the highest efficiency towards the desired output product, producing the highest yield of product while requiring the least amount of money to purchase and operate. Normal operating expenses include energy input, energy removal, raw material costs, labor, etc. Energy changes can come in the form of heating or cooling, pumping to increase pressure, frictional pressure loss, agitation, etc.

The chemical reactions take place inside a vessel we refer to as a reaction vessel.

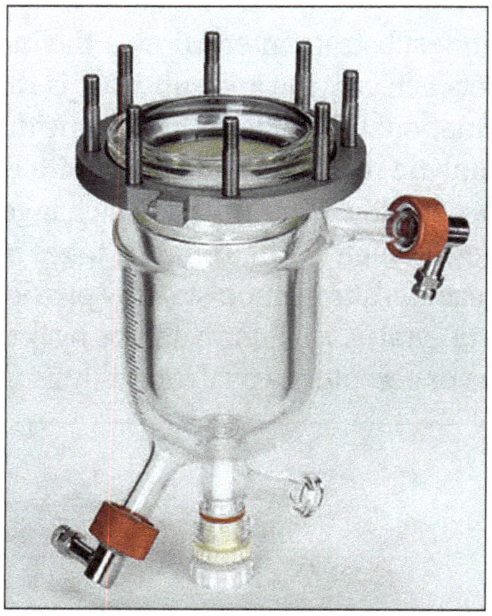

A reaction vessel: Notice there is actually a glass inside another and the other glass contain is used to allow cooling liquids to flow through.

Usually it is jacketed, meaning there is a sort of glass container around the main vessel in which oil, water or some other liquid is pumped through to allow cooling of the reaction vessel.

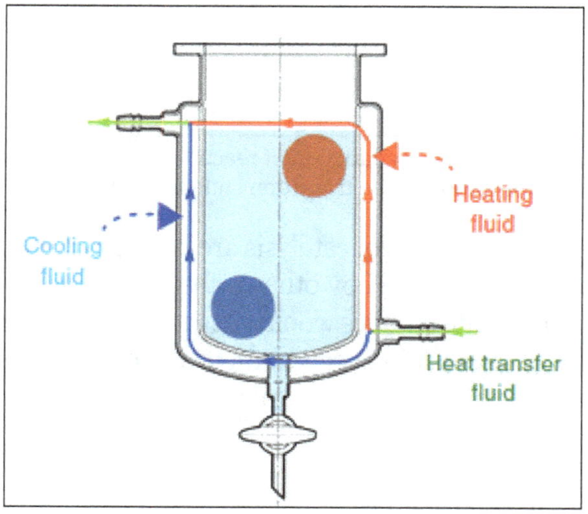

Illustration showing how there is a glass container outside the vessel which allows the flow of cooling liquids. Many chemical reactions require precise temperatures to happen optimally. Using heating with a bunsen burner and cooling liquids temperatures in the vessel can be adjusted up or down.

To mix everything properly, we have a motor driven *agitator* suspended in the mix of chemicals. It looks like a propeller.

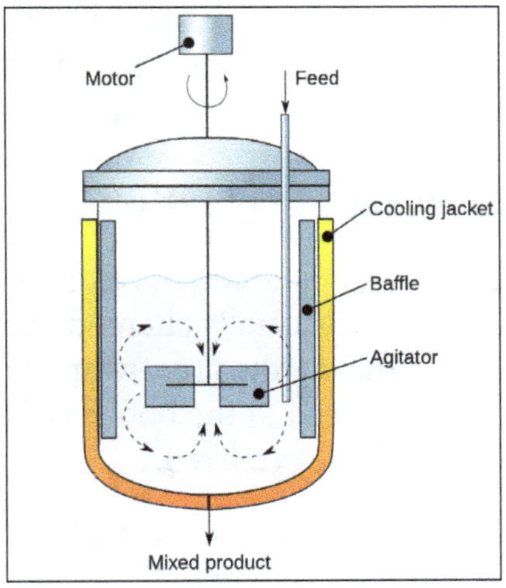

The vessel has a lid with many opening to attach different measuring instruments, or tubes for feedstock.

This is a flat flange lid with five necks.

Parts on the Reaction Vessel

- Hose coupling: The thing you screw or tighten to connect the hose to the vessel. You can refer to the *hose connection* as well.
- Vessel clamp: Ring around the vessel you use to attach it to the rest of the setup.
- Necks: The parts sticking up from the lid where you can connected tubes.
- Vessel support collar: Holds up the vessel.
- Quick release clamp: Clamp you can quickly open without using a screw driver, screwing etc. E.g. think about modern bike seats and old ones which you had to adjust with a hex or screwdriver.

Batch Reactor

A batch reactor is the simplest type of reactor vessels use for chemical or industrial processes. A typical batch reactor consists of a tank where chemical reactions occur. These tanks also have an agitator and an internal heating or cooling system. Tank sizes range from one liter to 15,000 liters.

Modeling Corrosion and Corrosion Protection

A batch reactor is commonly used in steel, stainless steel, glass or exotic alloy production. The liquids or solids used in the batch reactor are usually charged with connections in the reactor's top covers. Any vapors and gases generated during the process are discharged from connections in the top cover. Discharge outlets are usually placed at the bottom of the batch reactor to drain liquids that remain after the chemical reactions have occurred.

The versatility offered by a batch reactor is one of its best advantages, because several operations can be performed in a single vessel without needing to break the containment. This feature is useful when toxic or highly potent compounds are processed.

Batch reactor is standard equipment in chemical process industries. It consists essentially of an autoclave provided with mixing and heat transfer devices, as well as with piping connections for feeding the reactants and taking-off the products. The operation mode is intrinsically dynamic, which means that the properties vary continuously in time, but not in space because of homogenisation by mixing. For an isothermal operation, there is a perfect analogy between the evolution in time of concentrations and reaction rate in a batch reactor and the variation over length in an ideal continuous plug-flow reactor (PFR).

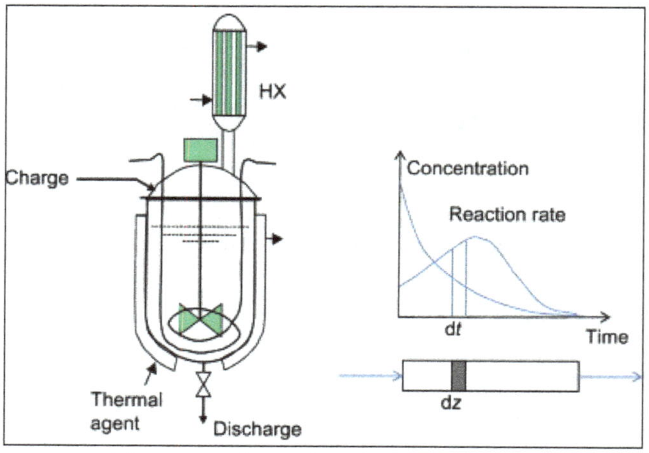

Batch reactor and the comparison of performance with a plug-flow reactor.

Batch reactors are typically employed in laboratory for investigating the feasibility for manufacturing chemical products, namely from the viewpoint of chemistry. Another important application is the formulation of the recipe of a future chemical product in term of composition and operation procedure. Among advantages of batch reactors one can cite flexibility for multi-product processes, and relatively low cost of equipment and maintenance. A remarkable feature of batch reactors is the possibility of using directly the research in laboratory for scaling-up the industrial process. If only the reaction time is examined, a batch reactor offers the best productivity, the same compared with an ideal PFR. However, with respect to the total manufacturing time, including auxiliary steps (charging, heating, cooling, discharging and cleaning), the productivity becomes much lower. Other disadvantages are higher manpower costs, more sophisticated process instrumentation and control, as well as the inherent variability of the product quality.

From operation viewpoint, one can distinguish between two modes:

- Fully batch process: reactants are initially charged and products discharged at the end.
- Semi-continuous process: at least one reactant or one product is removed continuously. This procedure is employed for manipulating the product distribution for equilibrium-limited reactions, for playing with kinetic effects for complex reactions and for avoiding runaway effects in the case of highly exothermic reactions.

Batch reactors are used in a number of industries producing small quantities of high-valued materials such as cell cultivation, polymer synthesis and crystallization. To operate the batch reactor with success, the final quality performance has to maintain with minimum cost. Though operated with the same recipe, the batch process shows batch-to-batch variations in its specified trajectories. Therefore, online process monitoring is essential to achieve successful batch operation.

Several techniques have been developed to deal with real-time observation of process variables and detection of abnormal operations. The model-based method is the most traditional way to estimate the operation trajectory. It consists of the first principle equation based on mass balance and empirical equations. Though this method shows considerable prediction performance, it is used in limited conditions because it needs much time and cost due to the complexity. Data-driven methods can be an alternative as they need little knowledge of the process. The stored data is used to extract correlation between process variables and quality indicators.

For batch processes, multiway principal component analysis (MPCA) was developed. In case both quality indicators and process variables are available, multiway partial least square (MPLS) was proposed to estimate batch performance. Since these models use samples from one operating point, they could deal with a static behaviour only. To analyze time dependent relations, dynamic PCA and dynamic PLS were developed. In this study, dynamic version of the MPLS method, called batch dynamic PLS (BDPLS), is used to construct the dynamic model.

Batch synchronization is an important pre-process because batch operations with a same recipe could have various batch lengths due to the inherent nonlinearity and operational constraints. Given the assumption that the data point at the same time point will be similar in each batch, several approaches are available to synchronize the sampled data for different batches. For example, indicator variables representing the batch progress for the reactions with well-known kinetics were proposed. After the reference batch is selected, the other batches are adjusted so that the indicator variables would have same values at the same time point.

Dynamic time warping (DTW) minimizes the distance between the trajectories of different time series, which can be used without any prior knowledge of the process. This study employed a phase-based DTW technique to synchronize the trajectories of different batches by adjusting each phase segment to be a same length.

Singular point (SP) refers to the point that has more information than others, for example, extrema. SPs can be used to separate the phase as the dynamic behavior changes at that point. Though the SP based phase division method has acceptable performance through the whole trajectory, it shows insufficient accuracy around the SP, when the phase changes. In this study, a new method for singular point based phase division is proposed by separating a transition phase from normal ones defined by the SP, so that the performance of the multiphase statistical model is improved.

Adiabatic Batch Reactor

An adiabatic batch reactor can be defined as a closed system with no input and output streams. This batch reactor training system is operated under adiabatic conditions (Reaction Mass Temperature is not kept constant and keeps on changing), perfect mixture (composition of reaction mixture is kept uniform throughout the operation) and constant volume (volume of the reaction mixture within the reactor is kept constant hence there is no measurable change in the density of reaction mass). This system is used to study catalytic homogeneous reaction under adiabatic condition.

Semi-batch Reactor

In a semi-batch reactor, some reactants are added to the reactor at the start of the batch, while others are fed continuously during the course of the reaction. Semi-batch reactors lie between batch and continuous reactors in terms of operation. A vertical reactor that can be configured to operate in semi-batch mode.

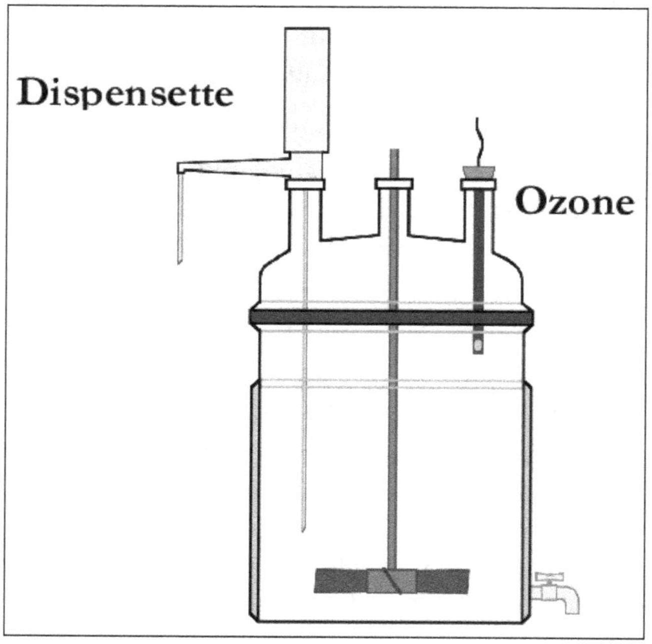

Semi-batch reactors occupy a middle ground between batch and continuous reactors. They are open systems like CSTRs and run on an unsteady-state basis like batch reactors. They usually consist of a single stirred tank, similar to a batch reactor. The half-pipe coil jacketed reactor shown below can be used in semi-batch operations.

Equipment Design

The movie below shows the operation of a semi-batch reactor. In this example, an initial amount of reactants is charged into the reactor. The reactor is then started, and additional reactants are added continuously to the tank. The reactor is then allowed to run until the desired conversion is achieved, at which point the products and remaining reactants are removed from the tank and the process can be started once more.

Examples: Semi-batch reactors are not used as often as other reactor types. However, they can be used for many two-phase (i.e. solid/liquid) reactions. Also, semi-batch reactors are used when a reaction has many unwanted side reactions, or has a high heat of reaction. By limiting the introduction of reactants, potential problems are eliminated.

The semi-batch slurry reactor pictured below is used in a lab for metallocene-catalyzed polymerization of alkenes.

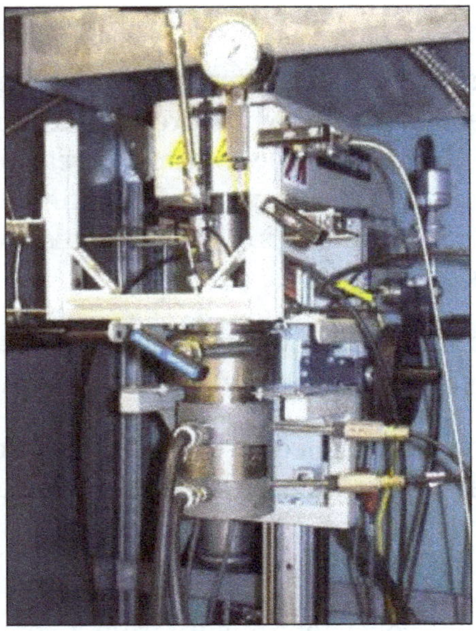

An exploded view of a semi-batch reactor is shown below, and is used to convert biomass feedstocks into usable products.

Advantages

- Good temperature control.
- Unwanted side-reactions minimized.

Disadvantages

- High labor cost per unit product.
- Large scale production difficult to achieve.
- Reactor operations difficult to analyse.

Continuous Reactor

An alternative to a batch process is to feed the reactants continuously into the reactor at one point, allow the reaction to take place and withdraw the products at another point. There must be an equal flow rate of reactants and products. While continuous reactors are rarely used in the laboratory, a water-softener can be regarded as an example of a continuous process. Hard water from the mains is passed through a tube containing an ion-exchange resin. Reaction occurs down the tube and soft water pours out at the exit.

Continuous reactors are normally installed when large quantities of a chemical are being produced. It is important that the reactor can operate for several months without a shutdown.

The residence time in the reactor is controlled by the feed rate of reactants to the reactor. For example, if a reactor has a volume of 20 m^3 and the feed rate of reactants is 40 $m^3\ h^{-1}$ the residence time is 20 m^3 /40 $m^3\ h^{-1}$ = 0.5 h. It is simple to control accurately the flow rate of reactants. The volume is fixed and therefore the residence time in the reactor is also well controlled.

The product tends to be of a more consistent quality from a continuous reactor because the reaction parameters (e.g. residence time, temperature and pressure) are better controlled than in batch operations.

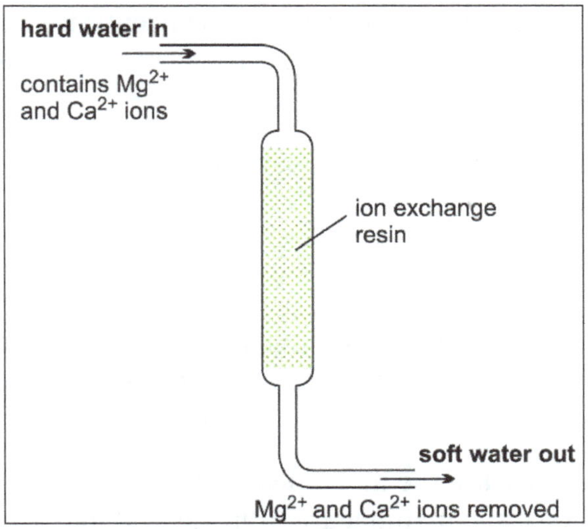

A continuous reactor.

They also produce less waste and require much lower storage of both raw materials and products resulting in a more efficient operation. Capital costs per tonne of product produced are consequently lower. The main disadvantage is their lack of flexibility as once the reactor has been built it is only in rare cases that it can be used to perform a different chemical reaction.

Types of Continuous Reactors

Industry uses several types of continuous reactors.

Tubular Reactors

In a tubular reactor, fluids (gases and/or liquids) flow through it at high velocities. As the reactants flow, for example along a heated pipe, they are converted to products. At these high velocities, the products are unable to diffuse back and there is little or no back mixing. The conditions are referred to as plug flow. This reduces the occurrence of side reactions and increases the yield of the desired product.

With a constant flow rate, the conditions at any one point remain constant with time and changes in time of the reaction are measured in terms of the position along the length of the tube.

The reaction rate is faster at the pipe inlet because the concentration of reactants is at its highest and the reaction rate reduces as the reactants flow through the pipe due to the decrease in concentration of the reactant.

Tubular reactors are used, for example, in the steam cracking of ethane, propane and butane and naphtha to produce alkenes.

Fixed Bed Reactors

A heterogeneous catalyst is used frequently in industry where gases flow through a solid catalyst (which is often in the form of small pellets to increase the surface area). It is often described as a fixed bed of catalyst.

Among the examples of their use are the manufacture of sulfuric acid (the Contact Process, with vanadium(V) oxide as catalyst), the manufacure of nitric acid and the manufacture of ammonia (the Haber Process, with iron as the catalyst).

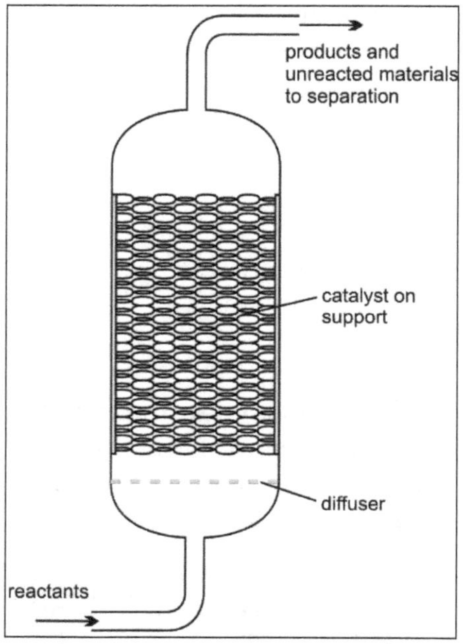

A fixed bed reactor.

A further example of a fixed bed reactor is in catalytic reforming of naphtha to produce branched chain alkanes, cycloalkanes and aromatic hydrocarbons using usually platinum or a platinum-rhenium alloy on an alumina support.

Fluid Bed Reactors

A fluid bed reactor is sometimes used whereby the catalyst particles, which are very fine, sit on a distributor plate. When the gaseous reactants pass through the distributor plate, the particles are carried with the gases forming a fluid. This ensures very good mixing of the reactants with the catalyst, with very high contact between the gaseous molecules and the catalyst and a good heat transfer. This results in a rapid reaction and a uniform mixture, reducing the variability of the process conditions.

One example of the use of fluid bed reactors is in the oxychlorination of ethene to chloroethene (vinyl chloride), the feedstock for the polymer poly(chloroethene) (PVC). The catalyst is copper(II) chloride and potassium chloride deposited on the surface of alumina. This support is so fine, it acts as a fluid when gases pass through it.

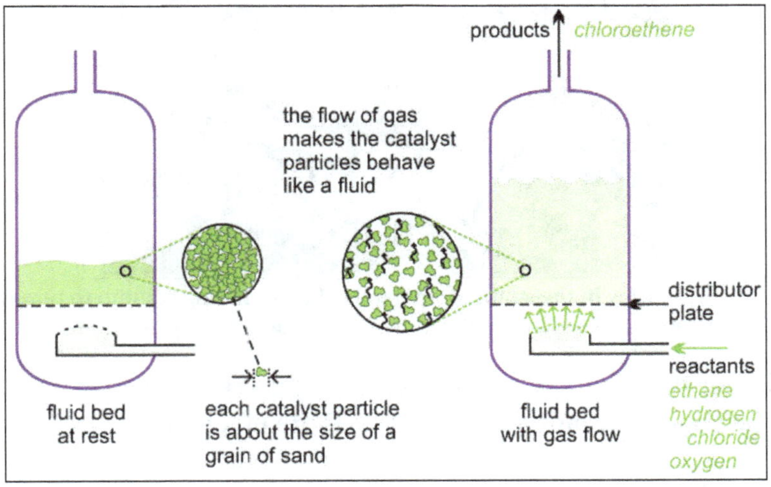

A diagram to illustrate a fluid bed reactor. On the left hand side, the particles are at rest. On the right hand side, the particles are now acting as a fluid, as the gaseous reactants pass through the solid.

Another example is the catalytic cracking of gas oil to produce alkenes (ethene and propene) and petrol with a high octane rating.

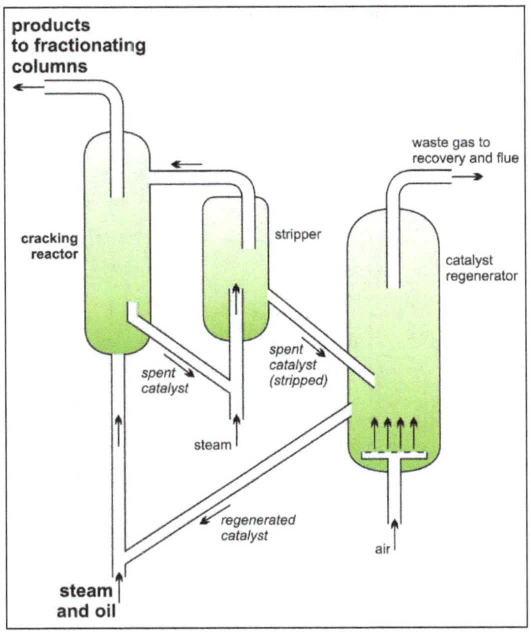

A catalytic cracker as used to produce alkenes from gas oil.

A silica-alumina catalyst is used but fine particles of carbon are rapidly deposited on its surface and the catalyst quickly becomes ineffective. The catalyst, still in the form of a fluid, is regenerated as it passes into a second vessel where it is heated strongly in air (sometimes with added oxygen) burn off the carbon and then returned into the reaction zone and mixed with gas oil.

Part of the catalytic cracker at Fawley in the south of England. 1 Fractionating column to remove and recover the butanes. 2 The catalyst regenerator. 3 Fractionating column to remove and recover ethane.

These reactors are larger than fixed bed reactors and are more expensive to construct. However, it is easier to control the conditions and the process is more efficient.

Continuous Stirred Tank Reactors

In a CSTR, one or more reactants, for example in solution or as a slurry, are introduced into a reactor equipped with an impeller (stirrer) and the products are removed continuously. The impeller stirs the reagents vigorously to ensure good mixing so that there is a uniform composition throughout. The composition at the outlet is the same as in the bulk in the reactor. These are exactly the opposite conditions to those in a tubular flow reactor where there is virtually no mixing of the reactants and the products.

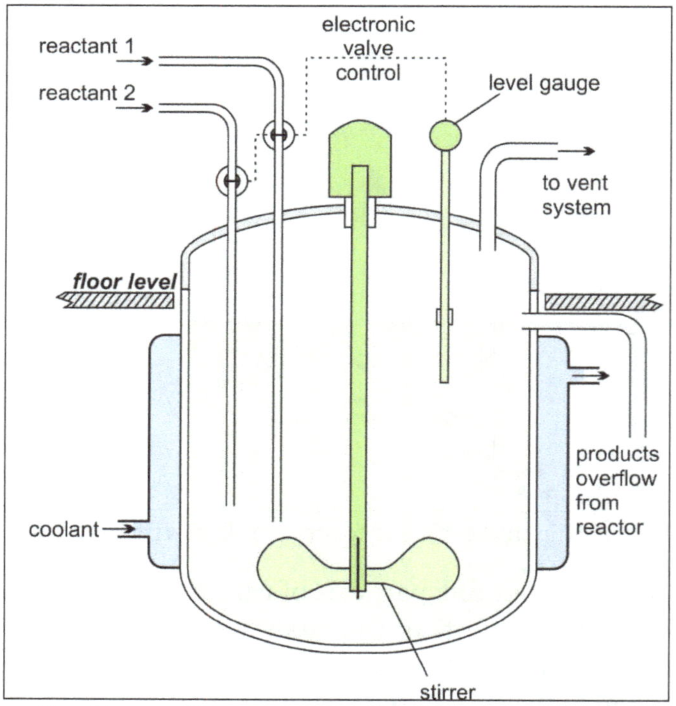

A line diagram illustrating a continuous stirred tank reactor.

A steady state must be reached where the flow rate into the reactor equals the flow rate out, for otherwise the tank would empty or overflow. The residence time is calculated by dividing the volume of the tank by the average volumetric flow rate. For example, if the flow of reactants is 10 $m^3 h^{-1}$ and the tank volume is 1 m^3, the residence time is 1/10 h, i.e. 6 minutes.

A CSTR reactor, used to make poly(ethene) in bulk.

A CSTR reactor is used, for example in the production of the amide intermediate formed in the process to produce methyl 2-methylpropenoate. Sulfuric acid and 2-hydroxy-2-methylpropanonitrile are fed into the tank at a temperature of 400 K. The heat generated by the reaction is removed by cooling water fed through coils and the residence time is about 15 minutes.

A variation of the CSTR is the loop reactor which is relatively simple and cheap to construct. In the diagram only one loop is shown. However, the residence time in the reactor is adjusted by altering the length or number of the loops in the reactor.

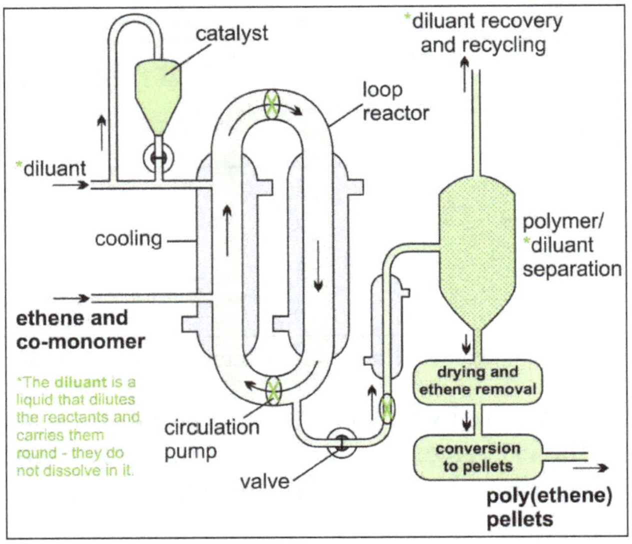

A line diagram showing a loop reactor used in the production of poly(ethene).

Loop reactors are used, for example, in the manufacture of poly(ethene) and the manufacture of poly(propene). Ethene (or propene) and the catalyst are mixed, under pressure, with a diluent, usually a hydrocarbon. A slurry is produced which is heated and circulated around the loops. Particles of the polymer gather at the bottom of one of the loop legs and, with some hydrocarbon diluent, are continuously released from the system. The diluent evaporates, leaving the solid polymer, and is then cooled to reform a liquid and passed back into the loop system, thus recirculating the hydrocarbon.

Heat Exchangers

Most chemical reactions are faster at higher temperatures and heat exchangers are frequently used to provide the heat necessary to increase the temperature of the reaction. A common heat exchanger is the shell and tube type where one part of the process flows through a tube and the other part around the shell.

A good example where heat exchange is important is in the manufacture of sulfur trioxide from sulfur dioxide in the contact process where the excess heat is used to warm incoming gases.

The heat from the reaction is transferred to incoming gases across the tube wall and the rate of heat transfer is proportional to:

- The temperature difference between the hot gases and the incoming gases.
- The total surface area of the tubes.

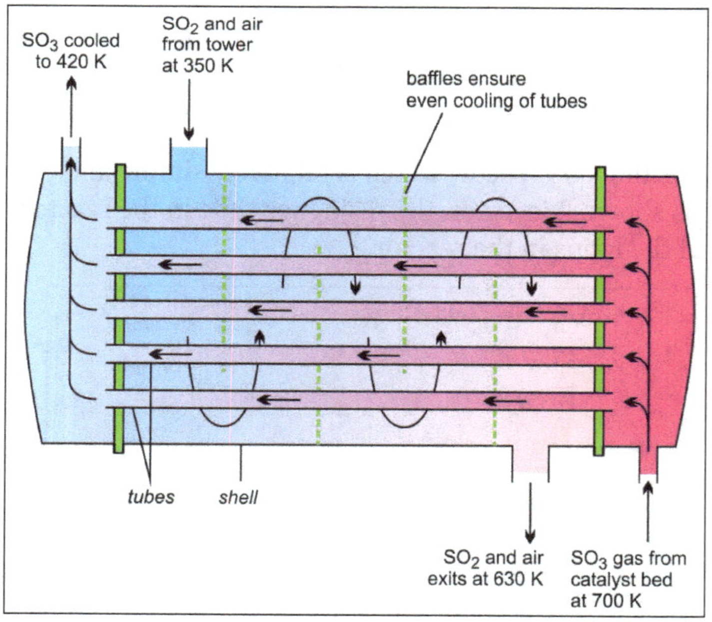

A heat exchanger used in the manufacture of sulfur trioxide.

Thus the rate of heat transfer required will determine the size of the exchanger but when a chemical reaction also occurs in the exchanger (as in the case of tubular reactors), it is important to take into account the residence time of the materials (whether they be gases or liquids) in the heat exchanger.

This exchanger is used for cooling product with water which enters the shell side at 1 and leaves at 2. The product to be cooled flows through the small tubes. The exchanger has been taken out of service to clean the tubes to improve the efficiency of the cooling process.

Another example of a heat exchanger is the condenser at the top of a distillation column. Here heat from the vapour emerging from the top of the column is removed by, for example, cooling water. The vapour cools down and condenses and the temperature of the water increases.

Some Developments

In the future many chemicals may be produced in reactors about the size of a large desktop computer, known as microreactors. The reduced size will lead to a reduction in capital costs and a reduction in the amount of chemicals in use at any one time, resulting in an inherently safer process.

The temperature can be kept constant more readily (as there is a much larger surface for a given volume). This allows for a very efficient heat transfer from the reaction to the surroundings even for very exothermic reactions such as the nitration of an aromatic hydrocarbon which can potentially be explosive.

There is considerable amount of research being carried out in developing microreactors. One example is the possibility of the direct conversion of benzene to phenol. A mixture of benzene and oxygen is fed through an alumina tube, packed with palladium at ca 350-400 K and hydrogen gas is passed over it.

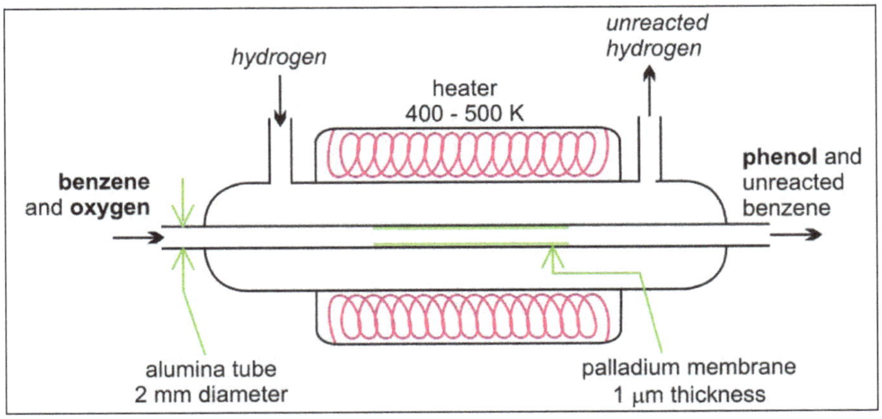

A microreactor being used to produce phenol from benzene.

Hydrogen permeates through the alumina tube, and is converted to atomic hydrogen by the palladium catalyst. The hydrogen atoms react with oxygen, releasing reactive oxygen species, such as hydroxyl radicals, which in turn react with the benzene to form phenol.

Another development is known as oscillatory flow mixing. Chemical engineers are designing reactors where the fluids to be reacted are oscillated inside a reactor with baffles at frequencies between 0.5 and 15 Hz with amplitudes in the range 1 to 100 mm. This allows for very effective mixing of the reactants and also for heat to be transferred to the surroundings. This gives similar conditions to those in plug flow which are otherwise difficult to achieve with small quantities of materials.

References

- Chemical-reaction, science: britannica.com, Retrieved 8 January, 2019
- "lab note #106 environmental impact of arc suppression". Arc suppression technologies. April 2011. Retrieved march 15, 2012
- Batch-reactor, definition: corrosionpedia.com, Retrieved 13 May, 2019
- Bendjebbar, andré (2000). Histoire secrète de la bombe atomique française. Documents (in french). Paris. Cherche midi. Isbn 978-2-862-74794-1
- Chemical-reactors, processes: essentialchemicalindustry.org, Retrieved 25 February, 2019
- Jogalekar, ashutosh. "leo szil rd, a traffic light and a slice of nuclear history". Scientific american. Retrieved 4 january 2016
- Chemical-kinetics, science: britannica.com, Retrieved 16 January, 2019
- Raven, peter h.; ray f. Evert; susan e. Eichhorn (2005). Biology of plants (7th ed.). New york: w.h. freeman and company publishers. Pp. 115–127. Isbn 0-7167-1007-2
- Chemical reactor, definition: definitions.net, Retrieved 30 April, 2019

4

Heat and Mass Transfer in Chemical Engineering

The study of the flow of heat in diverse process situations is known as heat transfer. A few ways in which transfer of heat takes place are conduction, convection and radiation. Mass transfer refers to the net movement of mass from one location to another. This chapter discusses in detail these ways of heat transfer as well as the coefficients of mass transfer.

Heat Transfer

Heat transfer is the study of the flow of heat. In chemical engineering, we have to know how to predict rates of heat transfer in a variety of process situations. For example, in mass transfer operations such as distillation, the overhead vapor has to be condensed to liquid product in a condenser, and the bottoms are boiled off into vapor in a reboiler. Often the feed stream is preheated using the bottoms product in a heat exchanger. Another example is the production and use of process steam, which is brought to various locations in a plant through steam pipes as a heating utility. Also, these steam pipes need to be insulated to minimize heat loss to the ambient air. Such insulation is also important when transporting hot fluids from one place to another. A similar application is the transport of refrigerated liquids through piping – here we need to insulate to avoid transferring heat into the liquid from the ambient air. Chemical reactors can generate heat if the reaction is exothermic, and this heat must be removed to avoid a runaway reaction; likewise, endothermic reactions need a supply of heat to maintain the reaction.

Heat transfer also is important in our daily lives. For example, we heat our homes in the winter using hot water in baseboard heaters. We boil water routinely for cooking purposes. If you look inside a modern personal computer, you'll see a fan that is used to cool the electrical circuitry, which becomes warm because of the flow of electrical current through resistances. Sometimes when the circuits are dense, a refrigerant is used in a sealed tube that is boiled at one end where it is warm, to take away the heat, and condensed at the other end where it is cooler.

The bodies of warm-blooded animals contain many examples of internal heat transfer as well as heat transfer with the surrounding air. Such animals possess a finely-tuned system that regulates their body temperature by adjusting rates of internal reactions to produce heat, varying blood flow rates as needed, and varying the diameter of blood vessels.

Finally, life on earth is possible only because of energy received from the sun by radiative heat transfer. Radiative heat transfer also is important in the manufacture of steel and other such materials, and in furnaces used for melting glass.

In all of these situations and many others, we can identify three basic mechanisms of heat transfer. They are conduction, convection, and radiation.

Conduction

Conduction occurs in stationary materials as a result of the vibrations of atoms or molecules in the materials. It is governed by Fourier's law of heat conduction, which in one dimension is written as:

$$Q_x = -kA \frac{\partial T}{\partial x} \text{ Btu/h or W}$$

Or

$$qx = \frac{Q_x}{A} = -k\frac{\partial T}{\partial x} \text{ Btu/h/ft}^2 \text{ or W/m}^2$$

Simply stated the heat flow per unit area is proportional to the negative of the temperature gradient. The proportionality constant is called the thermal conductivity, and it has units of Watts/meter/K or Btus/ft/°R. The thermal conductivities of typical materials vary widely by material, and it also depends on the temperature of the materials. There are different techniques available to solve the energy equation for conduction heat transfer: shell balance, conformal mapping, numerical methods, and graphical methods.

The more complete version of version of heat conduction equation can be written as below, where α is the thermal diffusivity, k is the thermal conductivity, and $\dot{q}$ is the heat flux:

$$\nabla^2 T + \frac{\dot{q}}{k} = \frac{1}{\alpha}\frac{\partial T}{\partial x}$$

The term $\nabla^2 T \equiv \nabla \cdot \nabla T$ is called the Laplacian. In Cartesian coordinate system, this is given as:

$$\nabla \cdot \nabla T = \left(\hat{i}\frac{\partial}{\partial x} + \hat{j}\frac{\partial}{\partial y} + \hat{k}\frac{\partial}{\partial z}\right) \cdot \left(\hat{i}\frac{\partial}{\partial x} + \hat{j}\frac{\partial}{\partial y} + \hat{k}\frac{\partial}{\partial z}\right)$$

Or

$$\nabla^2 T = \frac{\partial^2 T}{\partial x^2} + \frac{\partial^2 T}{\partial y^2} + \frac{\partial^2 T}{\partial z^2}$$

The Laplacian can also be expressed in cylindrical or spherical coordinates.

Basic Heat Conduction Equations

All three coordinates (i.e., Rectangular, Cylindrical, and Spherical) systems heat conduction equations in case of one-dimensional heat transfer is presented here, by expanding on equations above, we assume all the used parameters to be described as follows:

- Cp = specific heat of materials, J/(kg.°C).
- g = energy generation rate per unit volume, W/m^3
- q = conduction heat flux in the given coordinate direction, W/m^2

- t = time, s
- k = thermal conductivity of materials, where heat transferring through, W/(m °C)
- ρ = density of materials, kg/m³

Rectangular Coordinates

The heat transfer area A does not vary with variable x; hence, it is taken as constant and cancels. Then equations above reduce to:

$$\frac{\partial}{\partial x}\left(k\frac{\partial T\ (x,t)}{\partial x}\right)+g\ =\ \rho c_p\ \frac{\partial T\ (x,t)}{\partial x}$$

which is the one-dimensional, time-dependent heat conduction relation in the rectangular coordinate system.

Cylindrical Coordinates

In cylindrical coordinates, we represent the variable *x* with radial variable *r* thus equation above converts to a new form as:

$$\frac{1}{r}\frac{\partial}{\partial r}\left(rk\frac{\partial T\ (r,t)}{\partial r}\right)+g\ =\ \rho c_p\ \frac{\partial T\ (r,t)}{\partial t}$$

which is the one-dimensional, time-dependent heat conduction relation in the cylindrical coordinate system.

Spherical Coordinates

In spherical coordinates, we also represent the variable *x* with radial variable *r* thus equation above converts a new form as:

$$\frac{1}{r^2}\frac{\partial}{\partial r}\left(r^2k\frac{\partial T\ (r,t)}{\partial r}\right)+g\ =\ \rho c_p\ \frac{\partial T\ (r,t)}{\partial t}$$

which is the one-dimensional, time-dependent heat conduction relation in the spherical coordinate system.

A Compact Form of Basic Heat Conduction Equations

The compact version of equation $\frac{\partial}{\partial x}\ k\ \left(\frac{\partial T\ (x,t)}{\partial x}\right)+g\ =\ \rho c_p\ \frac{\partial T\ (x,t)}{\partial x}$ can be established using

Rectangular, Cylindrical, and Spherical coordinates systems for one-dimensional, time-dependent heat conduction as:

$$\frac{1}{r^n}\ \frac{\partial}{\partial r}\left(r^n\ k\ \frac{\partial T}{\partial r}\right)+g=\rho c_p\ \frac{\partial T}{\partial t}$$

where,

$$n = \begin{cases} 0 \text{ for ractangular coordinates} \\ 1 \text{ for cylindrical coordinates} \\ 2 \text{ for spherical coordinates} \end{cases}$$

and again remember in case of rectangular coordinate system the variable r should be replaced by variable x as a customary procedure.

Special Cases of Heat Conduction Equations

Several special cases of equation $\frac{1}{r^n}\frac{\partial}{\partial r}\left(r^n\, k\,\frac{\partial T}{\partial r}\right) + g = \rho c_p\,\frac{\partial T}{\partial t}$ are of practical interest in usage of thermal hydraulic subject. For constant thermal conductivity k, equation above reduces to a simplified form as:

$$\frac{1}{r^n}\frac{\partial}{\partial r}\left(r^n\,\frac{\partial T}{\partial r}\right) + \frac{1}{k}g = \frac{1}{\alpha}\frac{\partial T}{\partial t}$$

where,

$$\alpha \equiv \frac{k}{\rho c_p} = \text{thermal diffusivity of material, m}^2/\text{s}$$

For steady-state heat conduction with energy sources within the medium, equation above becomes:

$$\frac{1}{r^n}\frac{\partial}{\partial r}\left(r^n\, k\,\frac{\partial T}{\partial r}\right) + g = 0$$

and for the case of constant thermal conductivity, this result reduces to:

$$\frac{1}{r^n}\frac{\partial}{\partial r}\left(r^n\,\frac{\partial T}{\partial r}\right) + \frac{1}{k}g = 0$$

For steady-state heat conduction with no energy sources within the medium, equation $\frac{1}{r^n}\frac{\partial}{\partial r}\left(r^n\, k\,\frac{\partial T}{\partial r}\right) + g = \rho c_p\,\frac{\partial T}{\partial t}$, forms a more simplified version as:

$$\frac{d}{dr}\left(r^n k\frac{\partial T}{\partial r}\right) = 0$$

and for constant k, this result reduces to:

$$\frac{d}{dr}\left(r^n\,\frac{\partial T}{\partial r}\right) = 0$$

In all the equations of special cases of heat conduction, the exponent variable is defined as before:

$$n=\begin{cases}0 \text{ for ractangular coordinates}\\ 1 \text{ for cylindrical coordinates}\\ 2 \text{ for spherical coordinates}\end{cases}$$

And for rectangular coordinate again, Radial variable r is replaced by Cartesian variable x.

Heat Conduction in a Cylinder with a Uniform Heat Flux

This problem corresponds loosely to the generation of heat in a fuel rod in a nuclear reactor. From the second law of thermodynamics we know that the heat flows whenever there is a temperature difference, i.e., temperature gradient. The knowledge of the temperature distribution is essential to evaluate the heat flow. The temperature distribution and the heat flow constitute two basic elements in the design of thermal equipments such as boilers, heat exchangers, and nuclear reactor coresc.

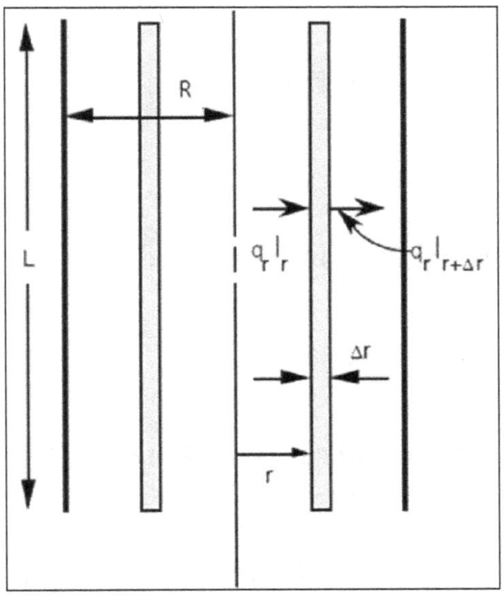

Shell energy balance for steady-state heat transfer in a rod with uniform heat generation.

Consider a cylinder with length L and outside radius r_o, as shown in figure. Look at a shell of thickness Δr and length L in the cylinder, and consider the various components, which contribute to the heat transport, in the r- direction:

- Flux of heat *in* across surface $r\left(\quad rLq\quad\right)\Big|$
- Flux of heat *out* across surface $r+\Delta r$ $\left(2\pi rLq_r\right)\Big|_{r+\Delta r}$
- Rate of heat production in shell volume $\left(2\pi r\Delta rL\right)\left(q''\right)$

Sum these contributions and set equal to zero. Divide by 2πΔrL and take the limit:

$$\lim_{\Delta r\to 0}\left(\frac{rq_r\big|_{r+\Delta r}+rq_r\big|_r}{\Delta r}\right)=q'''\,'r$$

The left side is the definition of the first derivative, $d(rq_r)/dr$. It can be integrated easily to give:

$$q_r = \frac{q'''r}{2} + \frac{C_1}{r}$$

Two boundary conditions exist to be satisfied, symmetry (limited heat flux) at $r = 0$, and $T = T_o$ at $r = r_o$ (one could also specify a boundary condition on heat flux, or the heat transfer coefficient). The symmetry boundary condition at $r = 0$ gives $C_1 = 0$.

Note what this equation. $\frac{1}{r^n}\frac{\partial}{\partial r}\left(r^n\frac{\partial T}{\partial r}\right) + \frac{1}{k}g = 0$ implies. If q''' is constant, the heat flux in the tube varies linearly from a maximum value at the wall to zero at the centerline. Now substituting Fourier's law for the heat flux term we get:

$$q_r = -k\frac{dT}{dr} = \frac{q'''r}{2}$$

If k and q''' are constant, integration again gives:

$$T = \frac{q'''r^2}{4k} + C_2$$

If the temperature is T_o at the wall, then C_2 is determined and

$$T - T_0 = \frac{q'''r_0^2}{4k}\left[1 - \left(\frac{r}{r_0}\right)^2\right]$$

This is a parabolic temperature profile and can be sketched as shown in figure.

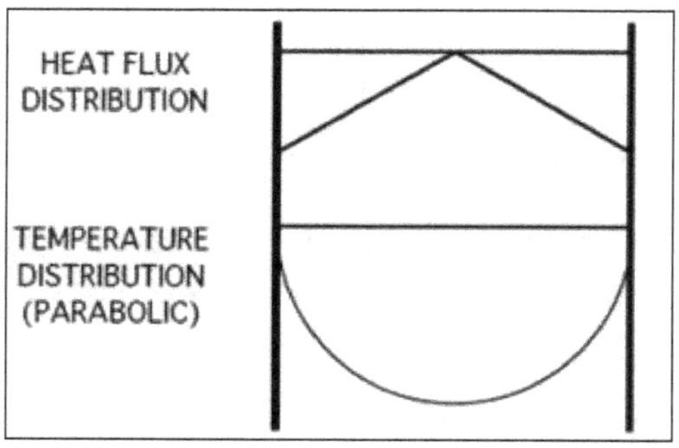

Heat flux and temperature distribution in a rod with uniform heat generation.

Heat Conduction in a Cylinder with a Uniform Heat Flux

For the case where a cylinder with heat generation is surrounded by a cladding material, as shown in figure, two additional boundary conditions are added at the interface between the two materials:

$$T_c = T\infty \quad at \quad r = r_c$$

$$q_f = q_c \text{ or } k_f \frac{dT_f}{dr} = k_c \frac{dT_c}{dr} dr \text{ at } r = r_f$$

The solution for the inner cylinder is already known. Now, it is desired to derive the solution for the cladding materials.

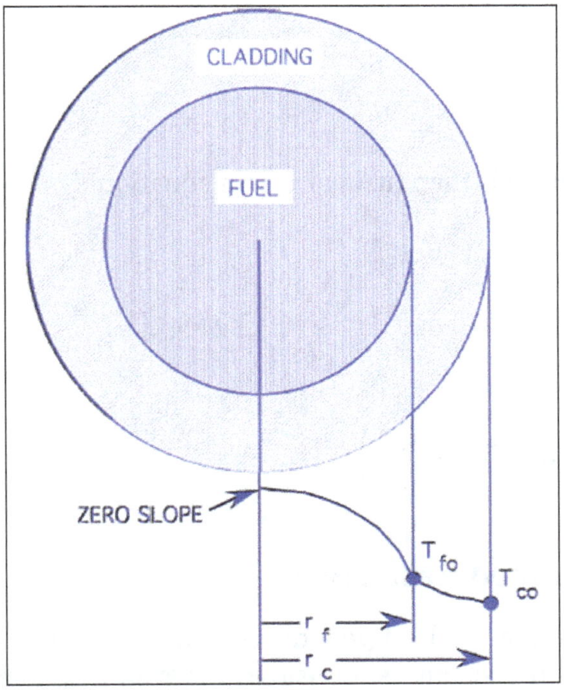

Temperature distribution in a rod with uniform heat generation with an outer jacket.

The solution for the jacket comes from the differential equation developed above:

$$\frac{d}{dr}(rq_r) = 0$$

This implies that the product of the heat flux with radius is constant, and equal to the heat flux at r = rf, which is known from the above problem to be:

$$rq_r = \frac{q''' r_f^2}{2}$$

So,

$$q_r = -K_c \frac{dT_c}{dr} = \frac{q''' r_f^2}{2r}$$

which can be integrated to give:

$$T_c = -\frac{q''' r_f^2}{2k_c}(\ln r + C)$$

The temperature at the cladding outer surface is known to be T_{co}, so,

$$T_c - T_{co} = \frac{q''' r_f^2}{2k_c} \ln\left(\frac{r_c}{r}\right)$$

Then, the temperature at the fuel/cladding interface is:

$$T_{fo} = T_{co} + \frac{q''' r_f^2}{2k_c} \ln\left(\frac{r_c}{r_f}\right)$$

and the maximum temperature is then (using $r = 0$ in equation $T = \frac{q''' r^2}{2} + C_2$)

$$T_{max} - T_{fo} = \frac{q''' r_f^2}{4k_f}$$

or

$$T_{max} = T_{co} + \frac{q''' r_f^2}{2}\left(\frac{1}{2k_f} + \frac{1}{K_c}\ln\left(\frac{r_c}{rf}\right)\right)$$

Composite Walls: Summed Resistance

For layers of materials (no internal heat generation), the total heat transport can be calculated using an analogy to electrical resistances. Consider, for example, a layered cylinder per demonstration in figure.

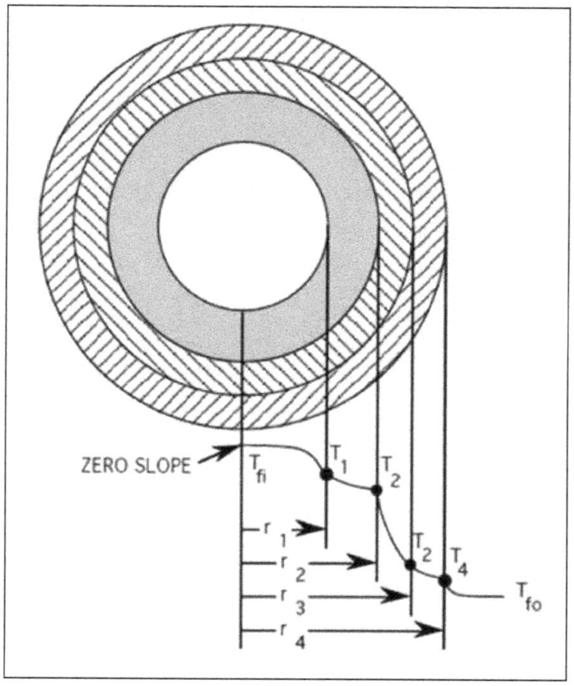

Heated rod with multiple jackets.

From equations $rq_r = \frac{q'''r_f^2}{2}$ and $q_r = K_c\frac{dT_c}{dr} = \frac{q'''r_f^2}{2r}$, we know that the heat transfer and temperature drop across any layer are related, $r_o q_o = r_1 q_1 = r_2 q_2 = \ldots$

$$T_2 - T_1 = \frac{r_0 q_0}{k_{12}} \ln\left(\frac{r_1}{r_2}\right) = \frac{q'}{2\pi L k_{12}} \ln\left(\frac{r_1}{r_2}\right)$$

where q' is the total heat transferred per unit length of the cylinder. We can write similar equations for the other layers:

$$T_3 - T_2 = \frac{q'}{2\pi L k_{23}} \ln\left(\frac{r_2}{r_3}\right)$$

In addition, the heat transfer resistance at the inner and outer fluid/solid interfaces can be expressed in terms of a heat transfer coefficient:

$$\begin{aligned} q''' = hA\Delta T &= h_i\,(2\pi r_1 L)(T_1 - T_{\mathrm{fi}}) \\ &= h_o\,(2\pi r_4 L)(T_{\mathrm{fo}} - T_4) \end{aligned}$$

All these expressions can be summed to eliminate the intermediate temperatures, giving:

$$q' = \frac{2\pi L(T_{\mathrm{fo}} - T_{\mathrm{fi}})}{\left(\frac{1}{r_1 h_i} + \frac{\ln(r_2/r_1)}{k_{12}} + \frac{\ln(r_3/r_2)}{k_{23}} + \frac{\ln(r_4/r_3)}{k_{34}} \frac{1}{r_4 h_o}\right)}$$

This can be written in a convenient form:

$$q' = \frac{(T_{\mathrm{fo}} - T_{\mathrm{fi}})}{(R_1 + R_2 + R_3 + R_4 + R_5)} = \frac{(T_{\mathrm{fo}} - T_{\mathrm{fi}})}{R_{\mathrm{Total}}}$$

where we define resistance R in the form of:

$$R_1 = \frac{1}{2\pi r_1 L h_i} \qquad R_2 = \frac{In(r_2/r_1)}{2\pi L k_{12}}$$

In general, such conduction resistances can be combined in series and parallel with the following addition rules:

$$R_{\mathrm{Total}} = \underbrace{\frac{1}{\frac{1}{R_1} + \frac{1}{R_2} + \frac{1}{R_3} + \ldots}}_{\text{parallel}} \qquad R_{\mathrm{Total}} = \underbrace{R_1 + R_2 + R_3 + \ldots}_{\text{Serial}}$$

Conduction in Complex Systems: Fuel Elements

In general, conduction problems are complicated by temperature and composition dependent thermal conductivities, in homogeneity from cracks, gaps, voids, and changing crystal structures, and nonuniform heat generation. The transport of heat in fuel elements is complicated by these entire phenomena.

Thermal Properties of Fuels

A variety of materials have been proposed and used as reactor fuels. Breeder reactors have used PuO_2 and UO_2, and interest is renewing in metal fuels for fast reactors. For thermal reactors, UO_2 has shown satisfactory chemical and irradiation tolerance, which overshadows the disadvantages of low thermal conductivity and uranium atom density. Table compares the oxide form of uranium to the metal, carbide, and nitride forms.

Table: Thermal properties of fuel materials.

Property	U	UO_2	UC	UN
Theoretical density at room temperature (kg/m³)	19.04×10³	10.97×10³	13.63×10³	14.32×10³
Metal density[a] (kg/m³)	19.04×10³	9.67×10³	12.97×10³	1360×10³
Melting point (°C)	1133	2800	2390	2800
Stability range	Up to 665 °C[b]	Up to m.p.	Up to m.p.	Up to m.p.
Thermal conductivity ave. 200–1000 °C(W/m°C)	32	3.6	23($UC_{1.1}$)	21
Linear coefficient of expansion		10.1×10⁻⁶	11.1×10⁻⁶	9.4×10⁻⁶
Tensile strength (MPa)	344–1380[c]	110	62	Not well defined

- [a]Uranium metal density in the compound at its theoretical density.
- [b]Addition of a small amount of Mo, Nb, Ti, or Zr extends stability up to the melting point.
- [c]The higher values apply to cold-worked metal.

The thermal conductivity of UO_2 at 95% of theoretical density, the thermal conductivity varies with temperature as:

$$k = \frac{3824}{402.4 + T} + 6.1256\times10^{-11}(T)^3$$

where T is Kelvin and k in watts per meter and Kelvin (W/m K). The thermal conductivity of UO_2 also changes with porosity, oxygen to metal atom ratio, PuO_2 content, cracking, and burnup.

Oxide fuel is manufactured by sintering pressed powdered UO_2 or mixed oxides at high temperature to create ceramic pellets. The pellets are deliberately manufactured with 5–10% porosity to prevent pellet swelling from gaseous fission product species. With a void fraction ϕ defined as the void volume divided by the total volume, Kampf and Karsten derived for negligible pore conductance a relationship between the thermal conductivity with zero voids, k_{TD} and with finite void fraction, k,

$$k = \left(1-\phi^{2/3}\right)k_{\mathrm{TD}}$$

The variation of the oxygen to metal ratio from the stoichiometric value of two decreases the thermal conductivity of UO_2. Likewise, the addition of PuO_2 results in a decrease in thermal conductivity of UO_2.

The cracking of fuel pellets introduces additional thermal resistance and thus decreases the effective thermal conductivity. Empirical relationships are available for predicting the change in effective thermal conductivity as a function of the fuel/ cladding gap dimensions under cold and hot conditions.

Fission gas release from the fuel is important, in that it degrades the relatively high thermal conductivity of the helium used to pressurize the fuel pin. The fraction of gas released from the pellets increases with increased operating temperature. The thermal conductivity of the gas mixture should be calculated using the Lindsay Bromley approach because a simple mole balance will under-predict the degrading effect of the higher molecular weight fission gases.

The gap between the fuel and cladding introduces a significant thermal resistance. It is normally accounted for using an effective heat transfer coefficient. For an open, non-contacting gap, the effective heat transfer coefficient is given by:

$$h_c = \frac{k_g}{\delta} + \sigma\left(\frac{1}{\varepsilon_s} + \frac{1}{\varepsilon_{sc}} - 1\right)\frac{T_s^4 - T_{sc}^4}{T_s - T_{sc}}$$

where k_g is the conductivity of the gas mixture, δ he gap thickness, σ the Stefan– Boltzmann constant, ε_s and ε_{sc} the emissivity of the fuel and cladding surfaces, and T_s and T_{sc} the surface temperatures of the fuel and cladding. Contact between the fuel and cladding increases the effective heat transfer coefficient, with the coefficient increasing with increasing contact pressure.

For a contacting gap, Majumdar and Tien have shown that the contact conductance due to contact is related to the load F as $h \propto F^{\eta D/2}$, where D is the fractal dimension for the surface and η is a variable ranging from 1 to 1.33 as a function of D. For machined surfaces, at small scales the fractal dimension takes on the Brownian value of 1.5, while at larger length scales the fractal dimension is larger due to the machining, approaching a value of 2.0. Due to the surface roughness, only a fraction of the total surface area comes in contact with flow. The gaps can then be viewed as creating a large number of parallel heat flow paths. From experimental data, heat transfer measurements for machined surfaces were found to be correlated with fractal dimensions varying from 1.7 to 1.99.

Under sustained operation at high power, the large temperature gradients cause migration of material from the pellet center, condensing in and filling voids at the outer radius. This results in void formation at the center of the pellets, and crystal structure changes in the fuel. For regions with temperatures above 1500 °C, the average grain size increases due to equiaxed grain growth. For temperatures above 1700 °C, columnar grains are formed in radial orientation within the fuel.

Other Problem in Heat Conduction

Heat conduction in fins: must set up energy balance, determine that:

$$\frac{d^2T}{dz^2} = \frac{hp}{kA}\left(T - T_{\infty}\right)$$

where h is the heat transfer coefficient, A the fin area, and P the fin perimeter.

Transient heat conduction to a semi-infinite slab has the same solution as that for flow near a wall suddenly set in motion. Solutions to the heat conduction equation (Laplace's equation) can also be solved by conformal mapping, such as Laplace Transformation or Fourier Transformation depends on boundary or initial conditions.

Convection

The second type of heat transfer to be examined is convection, where a key problem is determining the boundary conditions at a surface exposed to a flowing fluid.

Convection is the process of heat transfer by the bulk movement of molecules within fluids such as gases and liquids. The initial heat transfer between the object and the fluid takes place through conduction but the bulk heat transfer happens due to the motion of the fluid.

- Convection is the process of heat transfer in fluids by the actual motion of matter.
- It happens in liquids and gases.
- It may be natural or forced.
- It involves a bulk transfer of portions of the fluid.

When a fluid is heated from below, the thermal expansion takes place. The lower layers of the fluid, which are hotter, become less dense. We know that colder fluid is denser. Due to buoyancy, the less dense, hotter part of the fluid rises up. And the colder, denser fluid replaces it. This process is repeated when this part also gets heated and rises up to be replaced by the colder upper layer. This is how the heat is transferred through convection.

Convective heat transfer can be:

- Forced or assisted convection.
- Natural or free convection.
- Conductive Heat Transfer.

Forced or Assisted Convection

Forced convection occurs when a fluid flow is induced by an external force, such as a pump, fan or a mixer.

Natural or Free Convection

Natural convection is caused by buoyancy forces due to dens ity differences caused by temperature variations in the fluid. At heating the density change in the boundary layer will cause the fluid to rise and be replaced by cooler fluid that also will heat and rise. This continues phenomena is called free or natural convection. Boiling or condensing processes are also referred to as a convective heat transfer processes.

The heat transfer per unit surface through convection was first described by Newton and the relation is known as the Newton's Law of Cooling.

The equation for convection can be expressed as:

$$q = h_c A\, dT$$

where,

q = heat transferred per unit time (W, Btu/hr).

A = heat transfer area of the surface (m^2, ft^2).

h_c = convective heat transfer coefficient of the process ($W/(m^{2o}C$, $Btu/(ft^2\ h\ ^oF)$).

dT = temperature difference between the surface and the bulk fluid (^oC, F).

Heat Transfer Coefficients - Units

- 1 $W/(m^2K)$ = 0.85984 $kcal/(h\ m^2\ ^oC)$ = 0.1761 $Btu/(ft^2\ h\ ^oF)$
- 1 $Btu/(ft^2\ h\ ^oF)$ = 5.678 $W/(m^2\ K)$ = 4.882 $kcal/(h\ m^2\ ^oC)$
- 1 $kcal/(h\ m^2\ ^oC)$ = 1.163 $W/(m^2K)$ = 0.205 $Btu/(ft^2\ h\ ^oF)$

Convective Heat Transfer Coefficients

Convective heat transfer coefficients - h_c - depends on type of media, if its gas or liquid, and flow properties such as velocity, viscosity and other flow and temperature dependent properties.

Typical convective heat transfer coefficients for some common fluid flow applications:

- Free Convection - air, gases and dry vapors : 0.5 - 1000 ($W/(m^2K)$)
- Free Convection - water and liquids: 50 - 3000 ($W/(m^2K)$)
- Forced Convection - air, gases and dry vapors: 10 - 1000 ($W/(m^2K)$)
- Forced Convection - water and liquids: 50 - 10000 ($W/(m^2K)$)
- Forced Convection - liquid metals: 5000 - 40000 ($W/(m^2K)$)
- Boiling Water: 3.000 - 100.000 ($W/(m^2K)$)
- Condensing Water Vapor: 5.000 - 100.000 ($W/(m^2K)$)

Radiation

Radiation differs from Conduction and Convection heat transfer mechanisms, in the sense that it does not require the presence of a material medium to occur. Energy transfer by radiation occurs at the speed of light and suffers no attenuation in vacuum.

Radiation can occur between two bodies separated by a medium colder than both bodies.

According to Maxwell theory, energy transfer takes place via electromagnetic waves in radiation. Electromagnetic waves transport energy like other waves and travel at the speed of light.

Electromagnetic waves are characterized by their frequency ν (Hz) and wavelength λ (μm), where:

$$\lambda = c / \nu$$

where c is the speed of light in that medium; in a vacuum $c_o = 2.99 \times 10^8$ m/s. Note that the frequency and wavelength are inversely proportional. The speed of light in a medium is related to the speed of light in a vacuum,

$$c = c_o / n$$

where n is the index of refraction of the medium, n = 1 for air and n = 1.5 for water. Note that the frequency of an electromagnetic wave depends only on the source and is independent of the medium.

The frequency of an electromagnetic wave can range from a few cycles to millions of cycles and higher per second.

Einstein postulated another theory for electromagnetic radiation. Based on this theory, electromagnetic radiation is the propagation of a collection of discrete packets of energy called photons. In this view, each photon of frequency ν is considered to have energy of:

$$e = h\nu = hc / \lambda$$

where $h = 6.625 \times 10^{-34}$ J.s is the Planck's constant.

Note that in Einstein's theory h and c are constants, thus the energy of a photon is inversely proportional to its wavelength. Therefore, shorter wavelength radiation possesses more powerful photon energies (X-ray and gamma rays are highly destructive).

Electromagnetic radiation covers a wide range of wavelength, from 10^{-10} μm for cosmic rays to 10^{10} μm for electrical power waves.

Thermal radiation wave is a narrow band on the electromagnetic wave spectrum.

Thermal radiation emission is a direct result of vibrational and rotational motions of molecules, atoms, and electrons of a substance. Temperature is a measure of these activities. Thus, the rate of thermal radiation emission increases with increasing temperature.

What we call light is the visible portion of the electromagnetic spectrum which lies within the thermal radiation band.

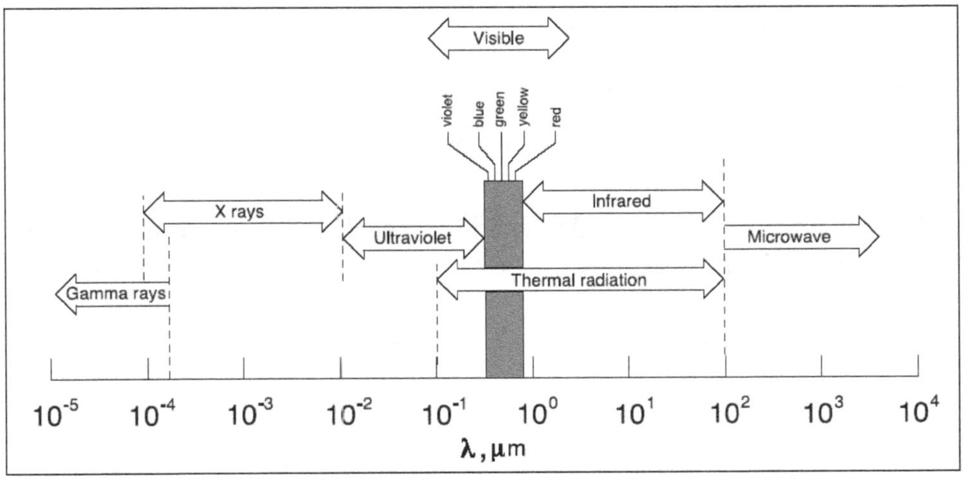

Electromagnetic spectrum.

Thermal radiation is a volumetric phenomenon. However, for opaque solids such as metals, radiation is considered to be a surface phenomenon, since the radiation emitted by the interior region never reach the surface.

Note that the radiation characteristics of surfaces can be changed completely by applying thin layers of coatings on them.

Blackbody Radiation

A blackbody is defined as a perfect emitter and absorber of radiation. At a specified temperature and wavelength, no surface can emit more energy than a blackbody. It is a diffuse emitter which means it emits radiation uniformly in all direction. Also a blackbody absorbs all incident radiation regardless of wavelength and direction.

The radiation energy emitted by a blackbody per unit time and per unit surface area can be determined from the Stefan-Boltzmann Law:

$$E_b = \sigma T^4 \left(w / m^2 \right)$$

where,

$$\sigma = 5.67 \times 10^{-8} \frac{w}{m^2 K^4}$$

where T is the absolute temperature of the surface in K and E_b is called the blackbody emissive power. A large cavity with a small opening closely resembles a blackbody.

Spectral blackbody emissive power is the amount of radiation energy emitted by a blackbody at an absolute temperature T per unit time, per unit surface area, and per unit wavelength.

$$E_{b\lambda}(T) = \frac{C_1}{\lambda^5 \left[\exp(C_2 / \lambda T) - 1 \right]} \left(\frac{w}{m^2 . \mu m} \right)$$

$$C_1 = 2\pi h c_0^2 = 3.742\times10^8 \left(W.\mu m^4 / m^2\right)$$
$$C_2 = hc_0 / k = 1.439\times10^4 \left(\mu m.K\right)$$
$$k = 1.3805\times10^{-23} \left(J / K\right) \text{Boltzmann's constant}$$

This is called Plank's distribution law and is valid for a surface in a vacuum or gas. For other mediums, it needs to be modified by replacing C_1 by C_1/n^2 , where n is the index of refraction of the medium,

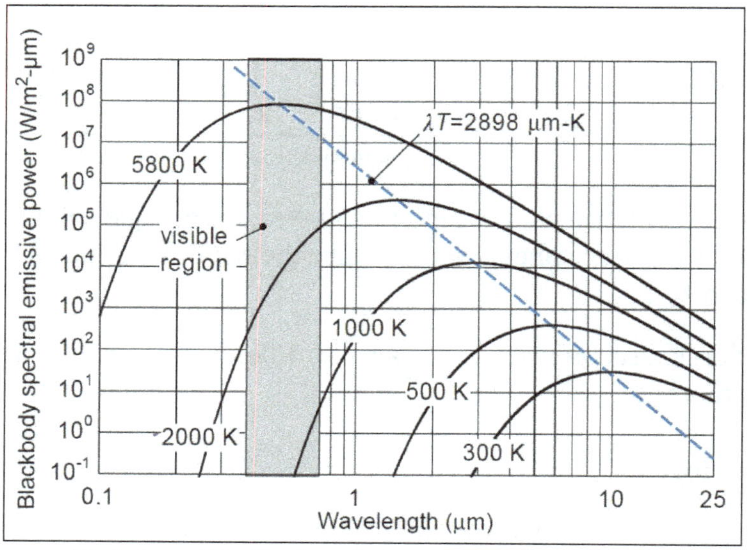

Variation of blackbody emissive power with wavelength.

The wavelength at which the peak emissive power occurs for a given temperature can be obtained from Wien's displacement law:

$$(\lambda T)_{\text{max power}} = 2897.8\ \mu m.K$$

It can be shown that integration of the spectral blackbody emissive power $E_{b\lambda}$ over the entire wavelength spectrum gives the total blackbody emissive power E_b:

$$E_b\ (T) = \int_0^{\infty} E_{b\lambda}(T)\, d\lambda\ =\ \sigma T^4\ \left(W / m^2\right)$$

The Stefan-Boltzmann law gives the total radiation emitted by a blackbody at all wavelengths from 0 to infinity. But, we are often interested in the amount of radiation emitted over some wavelength band.

To avoid numerical integration of the Planck's equation, a non-dimensional quantity $f\lambda$ is defined which is called the blackbody radiation function as:

$$f\lambda(T) \frac{\int_0^{\infty} E_{b\lambda}(T)\, d\lambda}{\sigma T^4}$$

The function f_λ represents the fraction of radiation emitted from a blackbody at temperature T in the wavelength band from 0 to λ.

Therefore, one can write:

$$f_{\lambda_1-\lambda_2}(T)=f_{\lambda_2}(T)-f_{\lambda_1}(T)$$
$$f_{\lambda-\infty}(T)=1-f_{\lambda}(T)$$

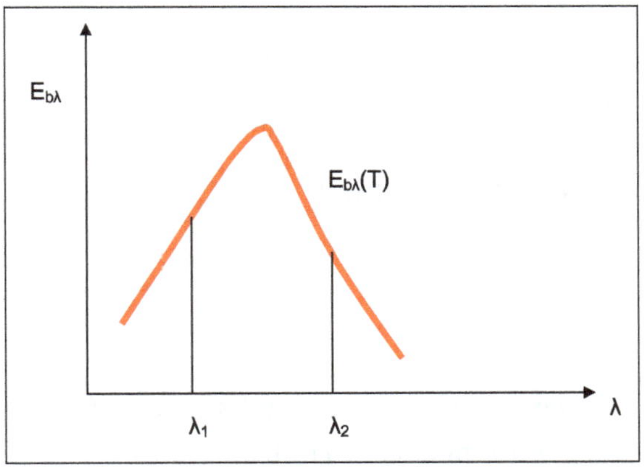

Fraction of radiation emitted in the wavelength between λ_1 and λ_2.

Example: The temperature of the filament of a light bulb is 2500 K. Assuming the filament to be a blackbody, determine the fraction of the radiant energy emitted by the filament that falls in the visible range. Also determine the wavelength at which the emission of radiation from the filament peaks.

Solution:

The visible range of the electromagnetic spectrum extends from 0.4 to 0.76 micro meter. Using table above:

$$\lambda_1 T=0.4\mu m(2500k)=1000\mu m.K \rightarrow f_{\lambda 1}=0.000321$$
$$\lambda_2 T=0.76\mu m(2500k)=1900\mu m.K \rightarrow f_{\lambda 2}=0.053035$$
$$f_{\lambda 2}-f_{\lambda 1}=0.05271$$

which means only about 5% of the radiation emitted by the filament of the light bulb falls in the visible range. The remaining 95% appears in the infrared region or the "invisible light".

Radiation Properties

A blackbody can serve as a convenient reference in describing the emission and absorption characteristics of real surfaces.

Emissivity

The emissivity of a surface is defined as the ratio of the radiation emitted by the surface to the radiation emitted by a blackbody at the same temperature. Thus,

$$0 \leq \varepsilon \leq 1$$

Emissivity is a measure of how closely a surface approximate a blackbody, $\varepsilon_{blackbody} = 1$.

The emissivity of a surface is not a constant; it is a function of temperature of the surface and wavelength and the direction of the emitted radiation, $\varepsilon = \varepsilon\ (T, \lambda, \theta)$ where θ is the angle between the direction and the normal of the surface.

The total emissivity of a surface is the average emissivity of a surface over all direction and wavelengths:

$$\varepsilon(T)=\frac{E(T)}{E_b(T)}=\frac{E(T)}{\sigma(T)^4}\rightarrow E(T)=\varepsilon(T)\sigma(T)^4$$

Spectral emissivity is defined in a similar manner:

$$\varepsilon_\lambda(T)\frac{E_\lambda(T)}{E_{b\lambda}(T)}$$

where $E_\lambda(T)$ is the spectral emissive power of the real surface. As shown, the radiation emission from a real surface differs from the Planck's distribution.

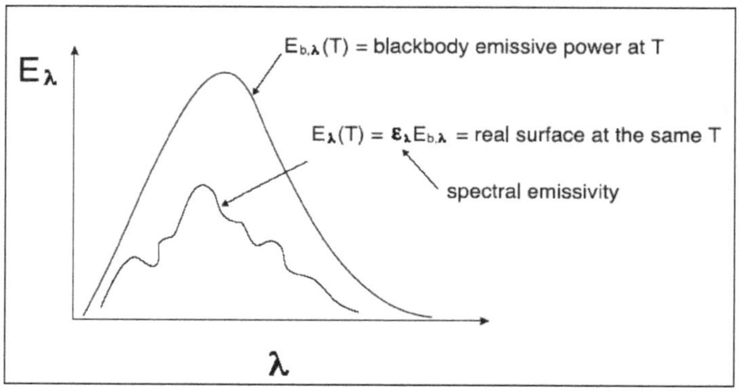

Comparison of the emissive power of a real surface and a blackbody.

To make the radiation calculations easier, we define the following approximations:

- Diffuse surface: is a surface which its properties are independent of direction.
- Gray surface: is a surface which its properties are independent from wavelength.

Therefore, the emissivity of a gray, diffuse surface is the total hemispherical (or simply the total) emissivity of that surface.

A gray surface should emit as much as radiation as the real surface it represents at the same temperature:

$$\varepsilon(T)=\frac{\int_0^\infty \varepsilon_\lambda(T)E_{b\lambda}(T)d\lambda}{\sigma T^4}$$

Absorptivity, Reflectivity and Transmissivity

The radiation energy incident on a surface per unit area per unit time is called irradiation, G.

- Absorptivity α: is the fraction of irradiation absorbed by the surface.
- Reflectivity ρ: is the fraction of irradiation reflected by the surface.
- Transmissivity τ: is the fraction of irradiation transmitted through the surface.
- Radiosity J: total radiation energy streaming from a surface, per unit area per unit time. It is the summation of the reflected and the emitted radiation.

$$\text{absorptivity}: \alpha = \frac{\text{absorbed radiation}}{\text{incident radiation}} = \frac{G_{abs}}{G} \quad 0 \le \alpha \le 1$$

$$\text{reflectivity}: \rho = \frac{\text{reflected radiation}}{\text{incident radiation}} = \frac{G_{ref}}{G} \quad 0 \le \rho \le 1$$

$$\text{transmissivity}: \tau = \frac{\text{transmitted radiation}}{\text{incident radiation}} = \frac{G_{tr}}{G} \quad 0 \le \tau \le 1$$

Applying the first law of thermodynamics, the sum of the absorbed, reflected, and the transmitted radiation radiations must be equal to the incident radiation:

$$G_{abs} + G_{ref} + G_{tr} = G$$

Divide by G:

$$\alpha + \rho + \tau = 1$$

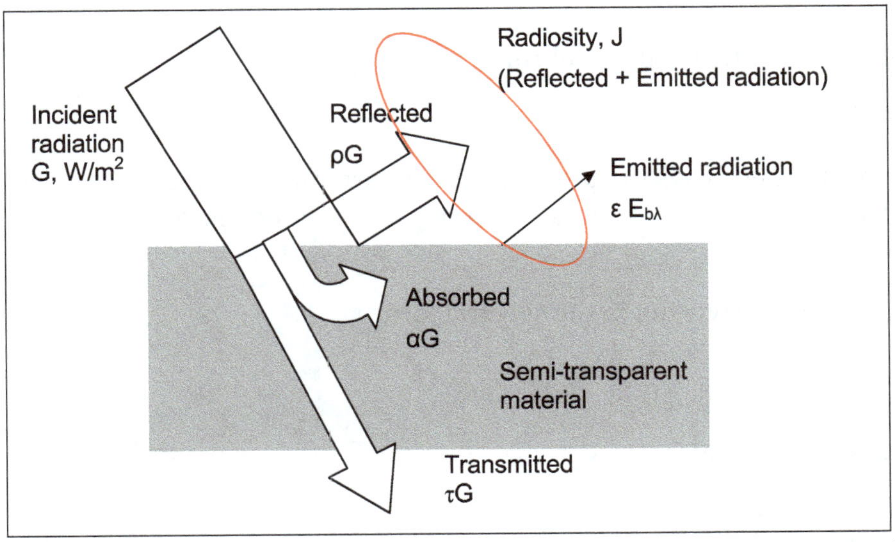

The absorption, reflection, and transmission of irradiation by a semitransparent material.

For opaque surfaces τ = 0 and thus: α + ρ = 1. The above definitions are for total hemi-spherical properties (over all direction and all frequencies). We can also define these properties in terms of their spectral counterparts:

$$G_\lambda + \rho_\lambda G_\lambda + \tau_\lambda G_\lambda = \alpha_\lambda G_\lambda$$

where,

$$\rho_\lambda = \rho\lambda(T,\lambda) \quad \text{spectral reflectivity}$$
$$\alpha_\lambda = \alpha_\lambda(T,\lambda) \quad \text{spectral absorptivity}$$
$$\tau_\lambda = \tau_\lambda(T,\lambda) \quad \text{spectral transmissivity}$$

thus,

$$1 = \rho_\lambda + \tau_A + \alpha_\lambda$$

Note that the absorptivity α is almost independent of surface temperature and it strongly depends on the temperature of the source at which the incident radiation is originating. For example α of the concrete roof is about 0.6 for solar radiation (source temperature 5762 K) and 0.9 for radiation originating from the surroundings (source temperature 300 K).

Kirchhoff's Law

Consider an isothermal cavity and a surface at the same temperature T. At the steady state (equilibrium) thermal condition:

$$G_{abs} = \alpha G = \alpha \sigma T^4$$

and radiation emitted:

$$E_{emit} = \varepsilon \sigma T^4$$

Since the small body is in thermal equilibrium, $G_{abs} = E_{emit}$

$$\varepsilon(T) = \alpha(T)$$

The total hemispherical emissivity of a surface at temperature T is equal to its total hemi-spherical absorptivity for radiation coming from a blackbody at the same temperature T. This is called the Kirchhoff's law.

The Kirchhoff's law can be written in the spectral form:

$$\varepsilon_\lambda(T) = \alpha_\lambda(T)$$

and in the spectral directional form:

$$\varepsilon_{\lambda,\theta}(T) = \alpha_{\lambda,\theta}(T)$$

The Kirchhoff's law makes the radiation analysis easier (ε = α), especially for opaque surfaces where ρ = 1 – α.

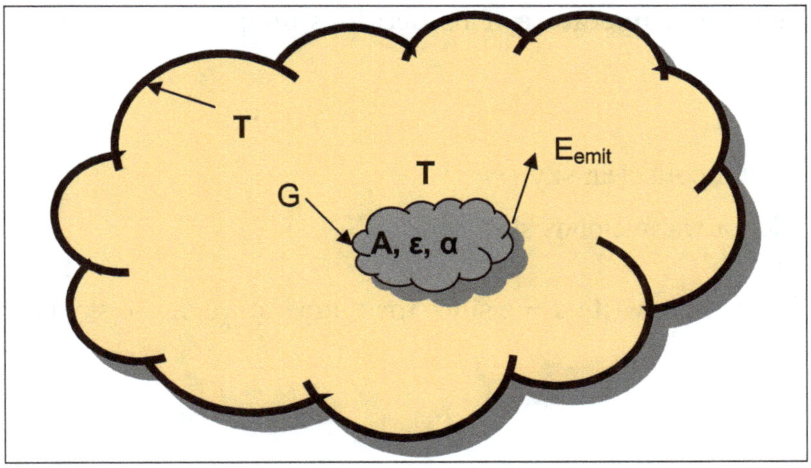

Small body contained in a large isothermal cavity.

Note that Kirchhoff's law cannot be used when there is a large temperature difference (more than 100 K) between the surface and the source temperature.

Solar Radiation

The solar energy reaching the edge of the earth's atmosphere is called the solar constant:

$$G_s = 1353 \text{ W} / \text{m}^2$$

Owing to the ellipticity of the earth's orbit, the actual solar constant changes throughout the year within +/- 3.4%. This variation is relatively small; thus G_s is assumed to be a constant.

The effective surface temperature of the sun can be estimated from the solar constant (by treating the sun as a blackbody).

The solar radiation undergoes considerable attenuation as it passes through the atmosphere as a result of absorption and scattering:

- Absorption by the oxygen occurs in a narrow band about $\lambda = 0.76$ μm.
- The ozone layer absorbs ultraviolet radiation at wavelengths below $\lambda = 0.3$ μm almost completely and radiation in the range of 0.3 – 0.4 μm considerably.
- Absorption in the infrared region is dominated by water vapor and carbon dioxide. Dust/pollutant particles also absorb radiation at various wavelengths.
- As a result the solar radiation reaching the earth's surface is about 950 W/m² on a clear day and much less on a cloudy day, in the wavelength band 0.3 to 2.5 μm.

Scattering and reflection by air molecules (and other particles) are other mechanisms that attenuate the solar radiation. Oxygen and nitrogen molecules scatter radiation at short wavelengths (corresponding to violet and blue colors). That is the reason the sky seems blue.

The gas molecules (mostly CO_2 and H_2O) and the suspended particles in the atmosphere emit radiation as well as absorbing it. It is convenient to consider the atmosphere (sky) as a blackbody at some lower temperature. This fictitious temperature is called the effective sky temperature T_{sky}.

$$G_{sky} = \sigma T^4_{sky}$$

$$T_{sky} = 230 \text{ K for cold clear sky}$$

$$T_{sky} = 285 \text{ K for warm cloudy sky}$$

Using Kirchhoff's law we can write α = ε since the temperature of the sky is on the order of the room temperature.

View Factor

Radiation heat transfer between surfaces depends on the orientation of the surfaces relative to each other as well as their radiation properties and temperatures.

View factor (or shape factor) is a purely geometrical parameter that accounts for the effects of orientation on radiation between surfaces.

In view factor calculations, we assume uniform radiation in all directions throughout the surface, i.e., surfaces are isothermal and diffuse. Also the medium between two surfaces does not absorb, emit, or scatter radiation.

$F_{i \to j}$ or F_{ij} = the fraction of the radiation leaving surface i that strikes surface j directly.

Note the following:

- The view factor ranges between zero and one.
- $F_{ij} = 0$ indicates that two surfaces do not see each other directly. $F_{ij} = 1$ indicates that the surface j completely surrounds surface i.
- The radiation that strikes a surface does not need to be absorbed by that surface.
- F_{ii} is the fraction of radiation leaving surface i that strikes itself directly. $F_{ii} = 0$ for plane or convex surfaces, and $F_{ii} \neq 0$ for concave surfaces.

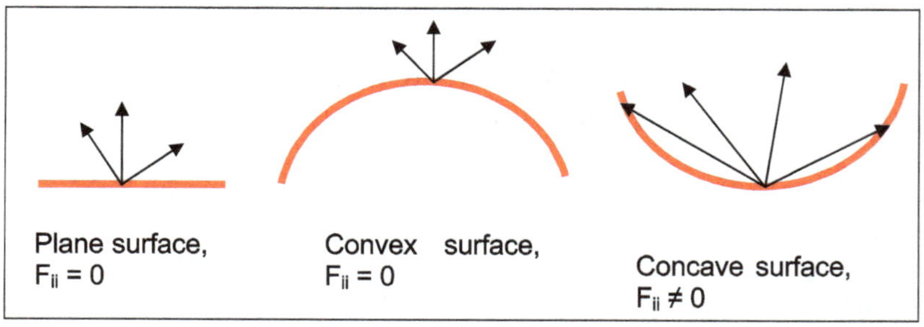

View factor between surface and itself.

View Factor Relations

Radiation analysis of an enclosure consisting of N surfaces requires the calculations of N^2 view factors. However, all of these calculations are not necessary. Once a sufficient number of view factors are available, the rest of them can be found using the following relations for view factors.

Reciprocity Rule

The view factor F_{ij} is not equal to F_{ji} unless the areas of the two surfaces are equal. It can be shown that:

$$A_i F_{ij} = A_j F_{ji}$$

Summation Rule

In radiation analysis, we usually form an enclosure. The conservation of energy principle requires that the entire radiation leaving any surface i of an enclosure be intercepted by the surfaces of enclosure. Therefore,

$$\sum_{j=1}^{N} F_{ij} = 1$$

The summation rule can be applied to each surface of an enclosure by varying i from 1 to N (number of surfaces). Thus the summation rule gives N equations. Also reciprocity rule gives 0.5 N (N-1) additional equations. Therefore, the total number of view factors that need to be evaluated directly for an N-surface enclosure becomes:

$$N^2 - \left[N + \frac{1}{2}N(N-1)\right] = \frac{1}{2}N(N-1)$$

Example: Determine the view factors F_{12} and F_{21} for the following geometries,

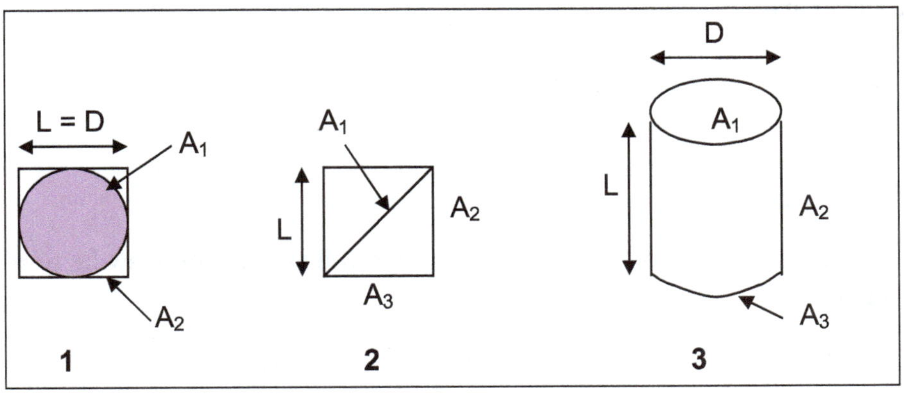

- Sphere of diameter D inside a cubical box of length L = D.
- Diagonal partition within a long square duct.
- End and side of a circular tube of equal length and diameter, L = D.

Assumptions: Diffuse surfaces.

Solution:

1) Sphere within a cube:

By inspection, $F_{12} = 1$

By reciprocity and summation:

$$F_{21} = \frac{A_1}{A_2} F_{12} = = \frac{\pi D^2}{6L^2} \times 1 = \frac{\pi}{6}$$

$$F_{21} + F_{22} = 1 \rightarrow F_{22} = 1 - \frac{\pi}{6}$$

2) Partition within a square duct:

From summation rule, $F_{11} + F_{12} + F_{13} = 1$ where $F_{11} = 0$

By symmetry $F_{12} = F_{13}$

Thus, $F_{12} = 0.5$.

From reciprocity:

$$F_{21} = \frac{A_1}{A_2} F_{12} = \frac{\sqrt{2L}}{L} \times 0.5 = 0.71$$

3) Circular tube: from figure with $r_2 / L = 0.5$ and $L / r_1 = 2$, $F_{13} \approx 0.17$. From summation rule,

$$F_{11} + F_{12} + F_{13} = 1 \quad \text{with } F_{11} = 0,\ F_{12} = 1 - F_{13} = 0.83$$

From reciprocity,

$$F_{21} = \frac{A_1}{A_2} F_{12} = \frac{\pi D^2 / 4}{\pi DL} \times 0.83 = 0.21$$

Superposition Rule

The view factor from a surface i to a surface j is equal to the sum of the view factors from surface i to the parts of surface j.

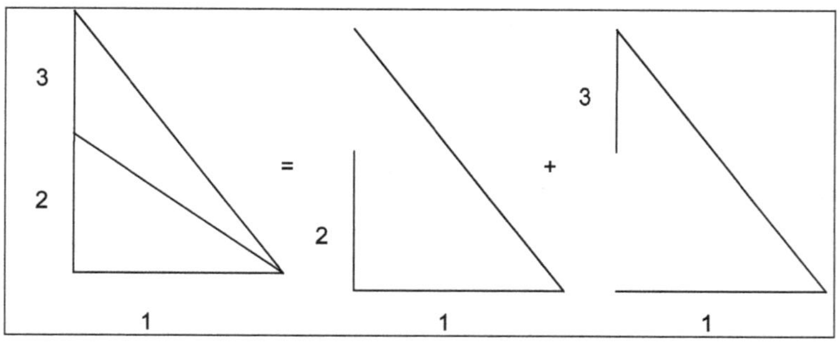

The superposition rule for view factors.

$$F_{1\to(2,3)} = F_{1\to 2} + F_{1\to 3}$$

Symmetry Rule

Two (or more) surfaces that possess symmetry about a third surface will have identical view factors from that surface.

Example: Find the view factor from the base of a pyramid to each of its four sides. The base is a square and its side surfaces are isosceles triangles.

From symmetry rule, we have:

$$F_{12} = F_{13} = F_{14} = F_{15}$$

Also, the summation rule yields: Also, the summation rule yields:

$$F_{11} + F_{12} + F_{13} + F_{14} + F_{15} = 1$$

Since, $F_{11} = 0$ (flat surface), we find; $F_{12} = F_{13} = F_{14} = F_{15} = 0.25$

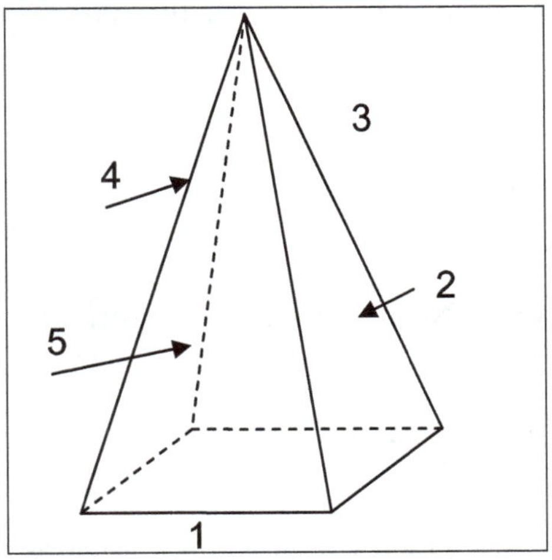

Crossed-strings Method

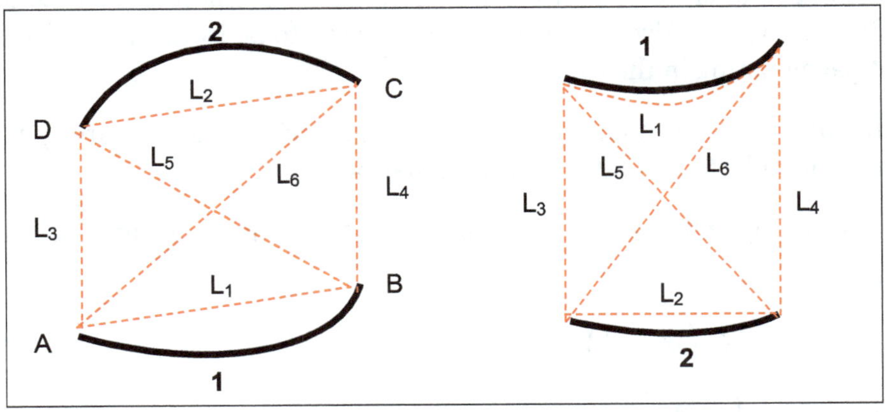

Cross-string method.

Geometries such as channels and ducts that are very long in one direction can be considered two-dimensional (since radiation through end surfaces can be neglected). The view factor between their surfaces can be determined by crossstring method developed by H. C. Hottel, as follows:

$$F_{i\to j} = \frac{\sum(\text{crossed strings}) - \sum(\text{uncrossed strings})}{2\times(\text{string on surface i})}$$

$$F_{12} = \frac{(L_5 + L_6) - (L_3 + L_4)}{2L_1}$$

Note that the surfaces do not need to be flat.

Radiation Heat Transfer

The analysis of radiation exchange between surfaces is complicated because of reflection. This can be simplified when surfaces are assumed to be black surfaces. The net radiation between two surfaces can be expressed as:

$$\dot{Q}_{12} = \begin{pmatrix}\text{radiation leaving surface 1}\\ \text{that directly stikes surface 2}\end{pmatrix} - \begin{pmatrix}\text{radiation leaving surface 2}\\ \text{that directly stikes surface 1}\end{pmatrix}$$

$$\dot{Q}_{12} = A_1\, F_{12}\, E_{b1} - A_2\, F_{21}\, E_{b2}\,(W)$$

Applying reciprocity $A_1\, F_{12} = A_2\, F_{21}$ yields:

$$\dot{Q}_{12} = A_1\, F_{12}\,\sigma\left(T_1^4 - T_2^4\right)(W)$$

Consider an enclosure consisting of N black surfaces maintained at specified temperatures. For each surface i, we can write:

$$\dot{Q}_i = \sum_{j=1}^{N} \dot{Q}_{ij} = \sum_{j=1}^{N} A_i\, F_{ij}\,\sigma\left(T_1^4 - T_2^4\right)(W)$$

Using the sign convention, a negative heat transfer rate indicates that the radiation heat transfer is to surface i (heat gain).

Now, we can extend this analysis to non-black surfaces. It is common to assume that the surfaces are opaque, diffuse, and gray. Also, surfaces are considered to be isothermal. Also the fluid inside the cavity is not participating in the radiation.

Radiosity J is the total radiation energy streaming from a surface, per unit area per unit time. It is the summation of the reflected and the emitted radiation.

For a surface i that is gray and opaque ($\varepsilon_i = \alpha_i$ and $\alpha_i + \rho_i = 1$), the Radiosity can be expressed as:

$$J_i = \varepsilon_i\, E_{bi} + \rho_i G_i$$

$$J_i = \varepsilon_i\, E_{bi} + (1 - \varepsilon_i) G_i \left(W / m^2\right)$$

$$J_i = \varepsilon_i\, E_{bi} = \sigma T_i^4 \quad (\text{for a blackbody})$$

Note that the radiosity of a blackbody is equal to its emissive power.

Using an energy balance, the net rate of radiation heat transfer from a surface i of surface area Ai can be expressed as:

$$\dot{Q}_i = A_i\left(j_i - G_i\right) \quad (W)$$

$$\dot{Q}_i = A_i\left(j_i - \frac{j_i - \varepsilon_i E_{bi}}{1-\varepsilon_i}\right) = \frac{A_i \varepsilon_i}{1-\varepsilon_i}\left(E_{bi} - j_i\right)$$

In electrical analogy to Ohm's law, a thermal resistance can be defined as:

$$\dot{Q}_i = \frac{E_{bi} - j_i}{R_i}$$

$$R_i = \frac{1-\varepsilon_i}{A_i \varepsilon_i}$$

where R_i is called the surface resistance to radiation.

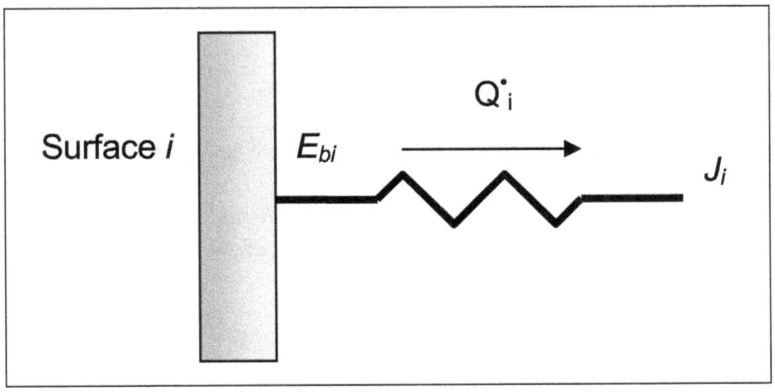

Surface resistance to radiation.

Note that the surface resistance to radiation for a blackbody is zero.

For insulated or adiabatic surfaces, the net heat transfer through them is zero. In this cases, the surface is called reradiating surface. There is no net heat transfer to a reradiating surface.

Net Radiation between Two Surfaces

Consider two diffuse, gray, and opaque surfaces of arbitrary shape maintained at uniform temperatures. The net rate of radiation heat transfer from surface i to surface j can be expressed:

$$\dot{Q}_i = A_i\, j_i F_{ij} - A_j\, j_j\, F_{ij} \quad (W)$$

Applying reciprocity,

$$\dot{Q}_i = A_i\, F_{ij}\left(j_i - j_j\right) \quad (W)$$

In analogy with Ohm's law, a resistance can be defined as:

$$\dot{Q}_i = \frac{(j_i - j_j)}{R_{ij}}$$

$$R_{ij} = \frac{1}{A_i F_{ij}}$$

where R_{ij} is called the space resistance to radiation.

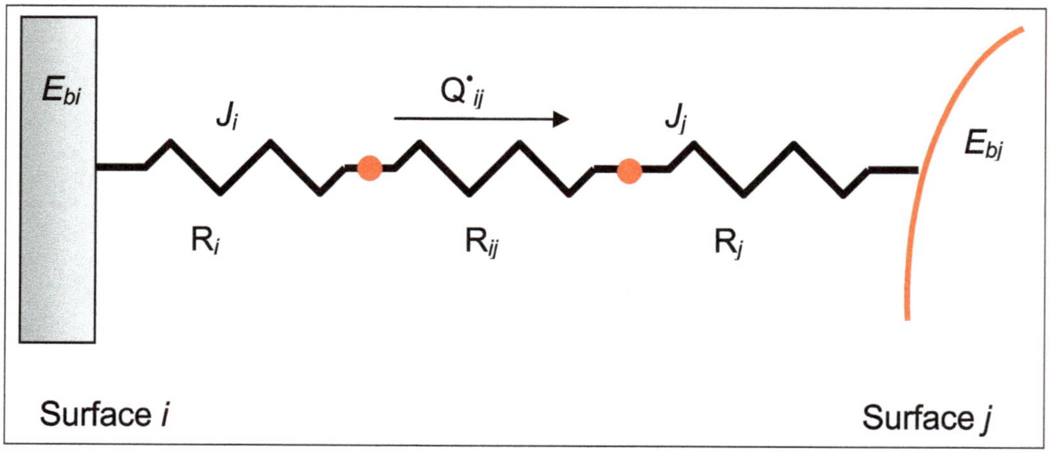

Electrical network, surface and space resistances.

In an N-surface enclosure, the conservation of energy principle requires that the net heat transfer from surface i to be equal to the sum of the net heat transfers from i to each of the N surfaces of the enclosure.

$$\dot{Q}_i = \sum_{j=1}^{N} \dot{Q}_{ij} = \sum_{j=1}^{N} \frac{j_i - j_j}{R_{ij}} \quad (W)$$

We have already derived a relationship for the net radiation from a surface:

$$\dot{Q}_i = \frac{E_{bi} - j_i}{R_i} \quad (W)$$

Combining these two relationships gives:

$$\frac{E_{bi} - j_i}{R_i} = \sum_{j=1}^{N} \frac{j_i - j_j}{R_{ij}}$$

Method of Solving Radiation Problem

In radiation problems, either the temperature or the net rate of heat transfer must be given for each of the surfaces to obtain a unique solution for the unknown surface temperature and heat transfer rates.

We use the network method which is based on the electrical network analogy. The following steps should be taken:

1. Form an enclosure; consider fictitious surface (s) for openings, room, etc.
2. Draw a surface resistance associated with each surface of the enclosure.
3. Connect the surface resistances with space resistances.
4. Solve the radiations problem (radiosities) by treating it as an electrical network problem.

Note that this method is not practical for enclosures with more than 4 surfaces.

Hot Plates in Room

Two parallel plates 0.5 by 1.0 m are spaced 0.5 m apart. One plate is maintained at 1000 °C and the other at 500 °C. The emissivities of the plates are 0.2 and 0.5, respectively. The plates are located in a very large room, the walls of which are maintained at 27 °C. The plates exchange heat with each other and with the room, but only the plate surfaces facing each other are to be considered in the analysis. Find the net heat transfer rate to each plate and the room; neglect other modes of heat transfer, i.e., conduction and convection.

Assumptions: Diffuse, gray, and opaque surfaces and steady-state heat transfer.

Solution: This is a three-body problem, the two plates and room. The radiation network is shown below.

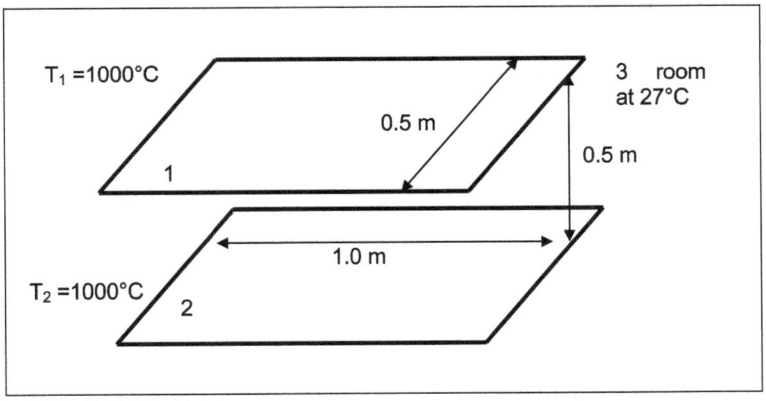

where,

$$T_1 = 1000\,°C = 1273\text{ K} \qquad A_1 = A_2 = 0.5\text{ m2}$$

$$T_2 = 500\,°C = 773\text{ K} \qquad T_3 = 27\,°C = 300\text{ K}$$

$$å_1 = 0.2 \qquad å_1 = 0.5$$

We can assume that the room is a blackbody, since its surface resistance is negligible:

$$R_3 = \frac{1-\varepsilon_3}{A_3\,\varepsilon_3} \approx 0 \quad A_3 \gg$$

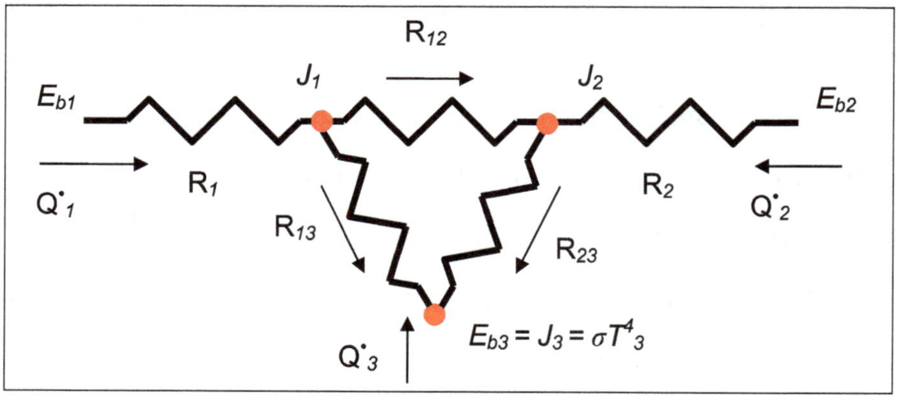

From figure, the shape factor F12 = 0.285

Using reciprocity and $A_1 = A_2$, $F_{12} = F_{21} = 0.285$

Applying summation rule: $F_{11} + F_{12} + F_{13} = 1$

Since $F_{11} = 0$ (flat plate), $F_{13} = 1 - 0.285 = 0.715$

Finally, from symmetry $F_{23} = F_{13} = 0.715$

The surface resistances are:

$$R_1 = \frac{1-\varepsilon_1}{A_1\,\varepsilon_1} = \frac{1-0.2}{(0.2)(0.5m^2)} = 8.0$$

$$R_2 = \frac{1-\varepsilon_2}{A_2\,\varepsilon_2} = \frac{1-0.5}{(0.5)(0.5m^2)} = 2.0$$

Space resistances are:

$$R_{12} = \frac{1}{A_1\,F_{12}} = \frac{1}{(0.285)(0.5m^2)} = 7.018$$

$$R_{23} = \frac{1}{A_2\,F_{23}} = \frac{1}{(0.715)(0.5m^2)} = 2.979$$

$$R_{13} = \frac{1}{A_1\,F_{13}} = \frac{1}{(0.175)(0.5m^2)} = 2.979$$

We need to find the radiosity for surface 1 and 2 only, since surface 3 is a blackbody, $J_3 = E_{b3} = \sigma T^4{}_3$

For node J_1:

$$\frac{E_{b1}-J_1}{R_1} + \frac{J_2-J_1}{R_{12}} + \frac{J_3-J_1}{R_{13}} = 0$$

For node J_2:

$$\frac{E_{b2}-J_2}{R_2}+\frac{J_1-J_2}{R_{12}}+\frac{J_3-J_2}{R_{23}}=0$$

where,

$$E_{b1}=\sigma T_1^4=148.87\,kW/m^2$$
$$E_{b2}=\sigma T_2^4=20.241\,kW/m^2$$
$$J_3=E_{b3}=\sigma T_3^4=0.4592\,kW/m^2$$

Substituting values and solving two equations, one finds:

$J_1 = 33.469\ kW/m^2$ and $J_2 = 15.054\ kW/m^2$

The total heat loss by plate 1 is:

$$\dot{Q}_1=\frac{E_{b1}-J_1}{R_1}=14.425\ kW$$

$$\dot{Q}_2=\frac{E_{b2}-J_2}{R_2}=2.594\ kW$$

The total radiation received by the room is:

$$\dot{Q}_3=\frac{J_1-J_3}{R_{13}}+\frac{J_2-J_3}{R_{23}}=17.020\ kW$$

Note that from an energy balance, we must have:

$$\dot{Q}_3=\dot{Q}_1+\dot{Q}_2$$

Mass Transfer

Mass transfer deals with situations in which there is more than one component present in a system; for instance, situations involving chemical reactions, dissolution, or mixing phenomena. A simple example of such a multicomponent system is a binary (two component) solution consisting of a solute in an excess of chemically different solvent.

Multicomponent Systems

In a multicomponent system, the velocity of different components is in general different. For example, in figure, pure gas A is present on the left and pure gas B on the right. When the wall separating the two gases is removed and the gases begin to mix, *A* will flow from left to right and B from right to left – clearly the velocities of *A* and *B* will be different.

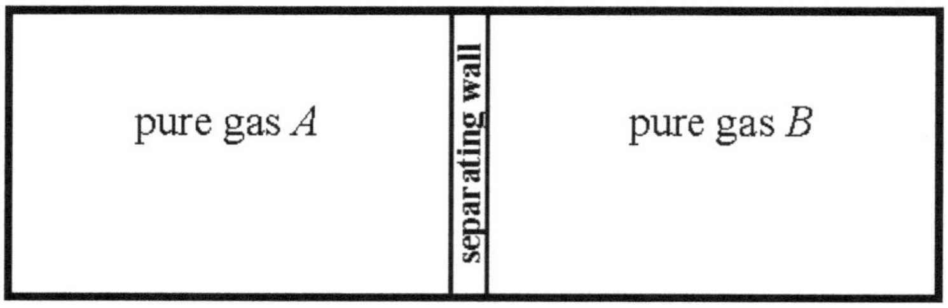

The velocity of particles (molecules) of component A (relative to the laboratory frame of reference) will be denoted v_A. Then, in this frame of reference, the molar flux N_A of species A (units: moles of A/(area time)) is:

$$N_A = c_A \, v_A$$

where c_A is the molar concentration of A (moles of A/volume). For example, equation above could be used to calculate how many moles of A flow through an area A_C per unit time. In figure, the flux is assumed to be normal to the area A_C. Then the amount of A carried across the area A_C per unit time is:

Amount of A carried through A_C per unit time = $N_A A_C = c_A \, v_A A_C$ (moles / time)

Since the volume swept out by the flow of A per unit time equals $v_A A_C$, the above expression is seen to equal this rate of volumetric "sweeping" times C_A, the amount of A per volume.

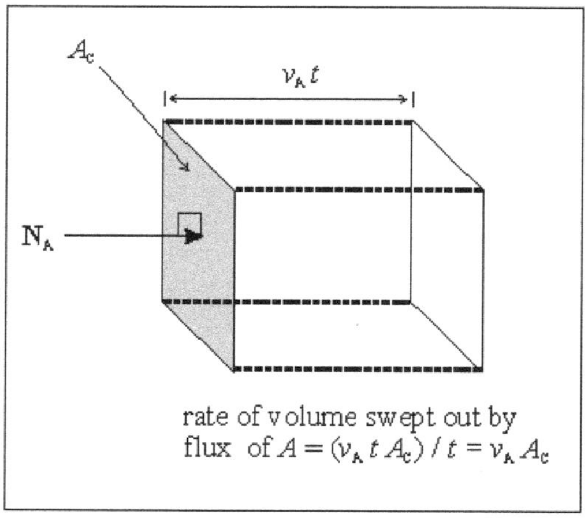

More generally, for arbitrary direction of N_A and a differential area element dB, the rate of A transport through dB would be,

$$\text{flux of A through dB} = -c_A v_A \cdot n \, dB \ \text{(moles / time)}$$

n is the outward unit normal vector to dB. One can understand equation above by realizing that $-v_A \cdot n \, dB$ is the volumetric flowrate of A species (volume/time) passing across dB from "outside" to "inside", where "outside" is pointed at by the unit normal vector n. Multiplying the volumetric flowrate $-v_A \cdot n \, dB$ by the number of moles of A per volume, c_A, equals the moles of A passing through dB per unit time.

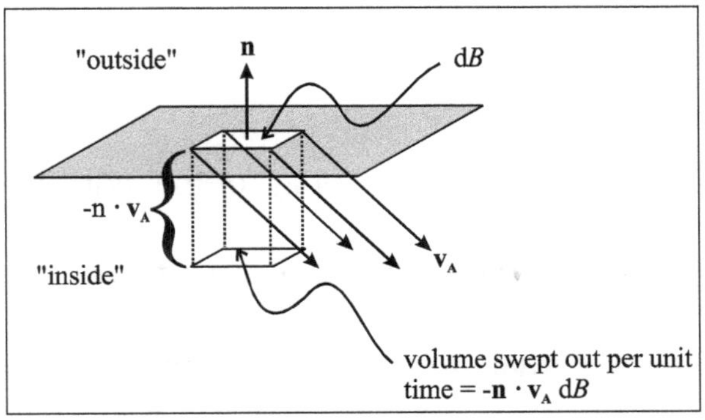

c_A is related to the total molar concentration c (c is moles of particles, irrespective of particle type, per volume) via:

$$c_A = x_A c$$

where x_A is the mole fraction of *A*. Summing over the mole fractions of all species must produce unity (*n* equals the total number of different species present in solution),

$$\sum_{i=l}^{n} x_i = 1$$

Similarly, we can also define a mass flux of *A*, n_A (units: mass of *A*/ (area time)),

$$n_A = \rho_A v_A$$

Here, v_A is still the velocity of species *A*, exactly the same as in equation $N_A = c_A v_A$. ρA is the mass concentration of A (mass of A per volume of solution),

$$\rho A = \omega_A \rho$$

where ρ is the total density (ρ is the summed mass of all particles, irrespective of particle type, per volume) of the solution and ω_A is called the mass fraction of A (i.e. $\omega_A = \rho_A/\rho$). Summing the mass fractions of all species must equal unity:

$$\sum_{i=l}^{n} \omega_i = 1$$

In general each chemical species "i" in a multicomponent mixture has a different velocity vi. However, it will nevertheless prove convenient to define an average velocity of the bulk fluid, a velocity that represents an average over all the v_i 's. In general, three types of average velocities are employed: mass average velocity v (v is what is usually dealt with in Fluid Mechanics), molar average velocity V, and volume average velocity v°. We will only deal with the first two average velocities, defined as follows:

$$v = \sum_{i=l}^{n} \omega_i v_i$$

$$V = \sum_{i=l}^{n} x_i \, V_i$$

From its definition, v is a mass fraction based average of the individual species' velocities, while V is a mole fraction based average. It can be shown that if total density $\rho = \sum_{i=l}^{n} \rho_i$ is constant irrespective of composition, and if the total molar concentration c = $\sum_{i=l}^{n} c_i$ is constant irrespective of composition, then v = V. It can further be shown that if all particles have the same mass m, so that $mi = m$ for all i where mi is mass of i type particles, then $x_i = \omega_i$ and therefore v = V.

The mass average velocity is what is needed in equations such as the Navier Stokes equations, which deal with momentum, a property that depends on how much mass is in motion. Thus, the amount of momentum per unit volume of a flowing multicomponent mixture is ρv (ρv = mv/*Volume*, where m is the total mass traveling with velocity v; m/Volume = ρ); thus momentum must be calculated using the mass average velocity v. Similarly, the equation of Continuity expresses conservation of mass, and is similarly written in terms of v. The physical laws expressed by these equations (conservation of momentum, conservation of mass) do not depend on the moles of particles involved, but they do depend on the mass of the particles.

On the other hand, when dealing with mass transfer, we will see that it is common to write some of the basic equations in terms of V as well as v. The reason for using V, in addition to v, is convenience. For instance, if in a particular problem there is no bulk flow of particles from one location to another so that, during the mass transfer process the number of particles at each point in space stays the same, then V = 0. Setting V to zero simplifies the mathematics. Imagine that, in their separated state as drawn, *A* and *B* are both ideal gases at the same pressure p and temperature T. Then, from the ideal gas equation, the molar concentration of *A* and *B* is the same,

$$c_A = c_B \;=\; c = \; p \,/\, RT \qquad (R = \text{gas contant})$$

The equality of C_A and C_B to the total concentration c is appropriate because the gases are pure; thus in each compartment the concentration of the gas (*A* or *B*) must also equal the total concentration c. After the separating wall is removed, particles of *A* and *B* will mix until a uniform composition is achieved throughout the vessel. In the final state, assuming the gases remain ideal when mixed, the value of p and T will remain the same as in the unmixed state and therefore the total concentration c also remains the same, $c = p/RT$ (p is now the total pressure, a sum of the partial pressures of *A* and *B*). Thus, in the final mixed state, the number of particles per volume c (here a sum of particles of *A* and B types) is the same as the number of particles per volume in the initial unmixed state. Thus mixing produced no net transfer of particles from one side of the vessel to the other, it only mixed the different particle types together. Under these conditions, when there is not net transfer of particles from one part of a system to another, V = 0.

In contrast, for the same mixing process, in general v will not be zero. For example, imagine that mass of *A* particles is twice as large as that of *B* particles. Then in the initial unmixed state the left hand side of the vessel (filled with *A*) contains more mass, and the density (mass/volume) of the gas *A* is higher than that of *B* even though its concentration (particles/volume) is the same. Once *A* and *B* mix, however, the density everywhere will become uniform. For this uniformity to be achieved mass must have been transferred from the A side to the B side; therefore, in contrast to the molar average velocity V, the mass average velocity v was not zero during the mixing process.

Integral and Differential Balances on Chemical Species

We will refer to the species under consideration as species *A*. Following a derivation that parallels that employed for the other conservation laws, the first step in the derivation of a conservation law on the amount of species A is to perform a balance for a closed control volume *V'*. *V'* is enclosed by a closed surface *B*.

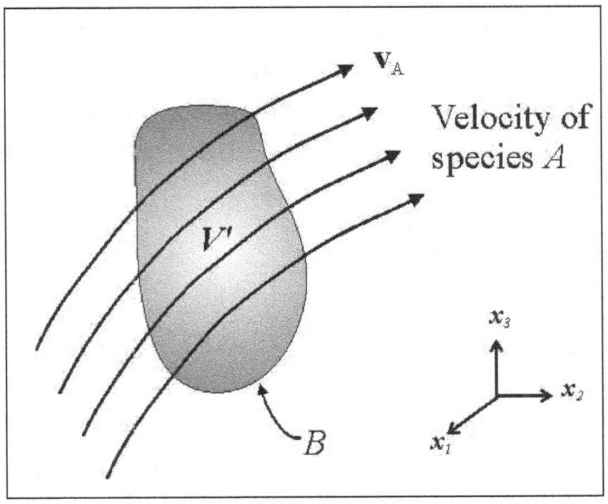

The amount of species A inside *V'* can change either due to convection through the boundary *B*, or by generation/consumption of *A* due to a chemical reaction. In words, the conservation for species A can stated as: The amount of species A inside *V'* can change either due to convection through the boundary *B*, or by generation/consumption of A due to a chemical reaction. In words, the conservation for species *A* can stated as:

Accumulation of *A* in *V'* = convection of *A* into *V'* + generation of *A* by chemical reactions.

An integral molar balance on species *A*, performed over the control volume *V'*, is written,

$$\frac{d}{dt}\left[\iiint_V c_A dV'\right] = -\iint_B c_A \mathrm{v}_A \cdot \mathrm{n\,dB} + \iiint_V R_A dV'$$

n is the outward unit normal vector to surface *B*, not to be confused with the mass flux $\mathrm{n_i} = \rho_i\ \mathrm{v_i}$ of species *i*. On the left side, c_A is concentration of A in moles per volume; thus $c_A dV'$ is the number of moles of *A* in a differential volume *dV'*. Integrating (i.e. summing) this term over the entire control volume *V'* yields the total number of moles of A in *V'*; the time derivative of this integral is the rate of change of moles of *A* inside *V'* (units: moles/time). Thus, the left hand term is just the rate of accumulation of *A* in *V'*, expressed in molar units.

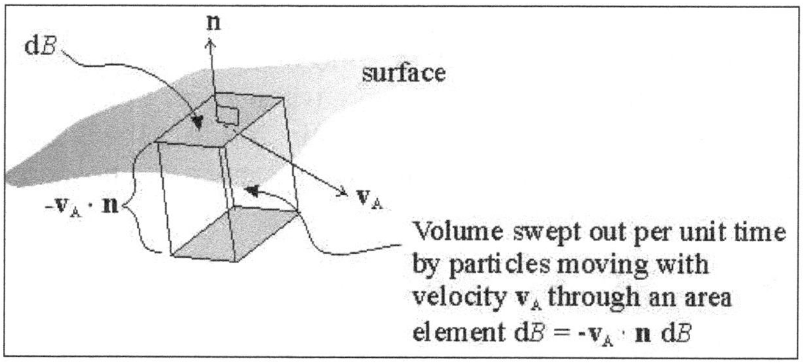

The accumulation term equals the rate at which *A* is convected into *V'* plus the rate at which *A* is generated inside *V'* by a homogeneous chemical reaction. The convection term can be understood by referring to figure $-\ v_A \cdot$ n is the component of the species velocity perpendicular to *B*, so that $-\ v_A \cdot$ nd *B* is the volumetric flowrate across the area element dB for particles traveling with a velocity v_A. Multiplying this volumetric flowrate by the moles of *A* per volume results in $-\ c_A v_A \cdot$ n dB, the molar flowrate of A through d *B*. Summing all the molar flowrates over the entire surface B then leads to the convection term (1st term on right) in equation ($\frac{d}{dt}\left[\iiint\limits_V c_A dV'\right] = -\iint\limits_B c_A v_A \cdot \mathrm{n\,dB} + \iiint\limits_V R_A dV'$).

The 2nd term on the right in equation ($\frac{d}{dt}\left[\iiint\limits_V C_A dV'\right] = -\iint\limits_B C_A V_A \cdot \mathrm{n\,dB} + \iiint\limits_V R_A dV'$) represents production of *A* by homogeneous reactions. A "homogeneous" reaction is one that occurs throughout the interior of *V'*. In contrast, a heterogeneous chemical reaction would be one that occurs only at an interface – for instance, between a solid and a liquid phase – and is not distributed throughout the entire volume. The molar reaction rate R_A has units of moles/(volume time) and represents the rate at which moles of *A* are produced or consumed by all homogeneous reactions. R_A dV' is the number of moles of *A* produced inside a volume element d*V'* per unit time (units: moles/time). Summing this production over the entire control volume leads to the total molar rate of production of *A*, inside *V'*, due to homogeneous chemical reactions.

Equation above, by assumption, did not include any generation of *A* due to heterogeneous reactions. Clearly, if in *V'* there was a large interface at which a heterogeneous reaction leads to production of *A*, one would have to add that term to equation above. The term would typically have the form of a rate of production of *A* per area (moles / (area time)) times the total area of the reacting surface. However, it may also be that a heterogeneous reaction is actually more conveniently modeled as homogeneous. For example, imagine that small catalyst particles (e.g. platinum powder) are suspended in a liquid inside *V'*, and that a reaction that produces *A* occurs on the surface of these powder particles. Because the reaction occurs only at the interface between a particle and the liquid, it is heterogeneous. However, since the particles are dispersed throughout *V'*, one could think of the reaction rate on a per volume basis (i.e. moles produced per volume of solution per time) as opposed to a per area basis (moles produced per surface area of particles per time).

As done previously for the other balances, one can: (1) use the Divergence Theorem to convert the surface integral of the convection term (1st term on right) into a volume integral, (2) move the d/dt derivative inside the accumulation integral since the integration limits are time independent (the limits do not depend on time because a fixed control volume is considered, whose shape and location do not change; this assumption can be relaxed at the expense of a somewhat more complicated mathematical expression), and (3) combine all terms under a common volume integral to obtain,

$$\iiint_V \left(\frac{\partial c_A}{\partial t} + \nabla \cdot c_A V_A - R_A\right) dV' = 0$$

The only way to ensure that equation ($\iiint_V \left(\frac{\partial c_A}{\partial t} + \nabla \cdot c_A V_A - R_A\right) dV' = 0$) evaluates to zero for an arbitrary control volume *V'* is to require that:

$$\frac{\partial c_A}{\partial t} + \nabla \cdot c_A \mathrm{v}_A - R_A = 0$$

$$\frac{\partial c_A}{\partial t} + \nabla \cdot c_A \mathrm{v}_A + R_A$$

Equation ($\frac{\partial c_A}{\partial t} + \nabla \cdot c_A \mathrm{V}_A + R_A$) is the differential molar balance on species *A*. It states that the rate of accumulation of moles of *A* at a point in space (left hand side) equals the rate at which moles of *A* are convected into that point (1st term on right), plus the rate at which moles of *A* are produced at that point by chemical reactions (2nd term on right). These physical interpretations can be verified by tracing the origin of the terms back to the corresponding terms in the integral balance, equation ($\frac{d}{dt}\left[\iiint_V C_A dV'\right] = -\iint_B C_A V_A \cdot \mathrm{n\,dB} + \iiint_V R_A dV'$).

Multiplication of equation ($\frac{d}{dt}\left[\iiint_V C_A dV'\right] = -\iint_B C_A V_A \cdot \mathrm{n\,dB} + \iiint_V R_A dV'$) by the molar mass M_A (mass/mole of *A*) of species *A*, and recognizing that ρ_A, the mass of *A* per volume, is given by:

$$\rho_A = M_A c_A$$

leads to the integral mass balance on species *A*,

$$\frac{d}{dt}\left[\iiint_V \rho_A dV'\right] = -\iint_B \rho_A \mathrm{V}_A \cdot \mathrm{n\,dB} + \iiint_V r_A dV'$$

In equation above, the mass reaction rate r_A has units of mass/(volume time) and represents the production or consumption of mass of species *A* by all chemical reactions. r_A is given by:

$$r_A = M_A R_A$$

Through manipulations analogous to those that led to ($\frac{\partial c_A}{\partial t} + \nabla \cdot c_A \mathrm{V}_A + R_A$), equation ($\frac{d}{dt}\left[\iiint_V \rho_A dV'\right] = -\iint_B \rho_A \mathrm{V}_A \cdot \mathrm{n\,dB} + \iiint_V r_A dV'$) can be converted to a differential mass balance on species A,

$$\frac{\partial \rho_A}{\partial t} = \nabla \cdot \rho_A \mathrm{v}_A + r_A$$

Recalling that:

$$\mathrm{N}_A = c_A \, \mathrm{v}_A \quad (\text{molar flux})$$

$$\mathrm{n}_A = \rho_A \, \mathrm{v}_A \quad (\text{mass flux})$$

Equations ($\frac{\partial c_A}{\partial t} + \nabla \cdot c_A \mathrm{v}_A + R_A$) and ($\frac{\partial \rho_A}{\partial t} + \nabla \cdot \rho_A \mathrm{v}_A + r_A$) can be written as:

$$\frac{\partial c_A}{\partial t} = -\nabla \cdot \mathrm{N}_A + R_A$$

$$\frac{\partial \rho_A}{\partial t} = -\nabla \cdot \mathrm{n}_A + r_A$$

Note that, in deriving these equations, no assumptions were made as to which component of a solution (i.e. a solute, the solvent, etc.) is represented as species *A* – therefore, these equations apply to each solute species as well as the solvent. Thus, if one chooses to label the solvent as species *A*, then a solute species could be labeled as species *B*. The equations that would be used for the solute *B* are exactly as in above equations and except that the subscript *A* would be replaced by the subscript *B*.

The differential species' balances were derived independent of any particular coordinate system. To apply them to solving a particular problem, one must first choose a coordinate system suited to describing the problem and then transcribe the equations into that coordinate system. For example, equation ($\frac{\partial \rho_A}{\partial t} + \nabla \cdot \rho_A \mathrm{v}_A + r_A$) becomes:

$$\frac{\partial \rho_A}{\partial t} = -\frac{\partial(\rho_A v_{A1})}{\partial x_1} - \frac{\partial(\rho_A v_{A2})}{\partial x_2} - \frac{\partial(\rho_A v_{A3})}{\partial x_3} + r_A \quad (\text{Cartesian "CCS" coordinates})$$

$$\frac{\partial \rho_A}{\partial t} = -\frac{1}{r}\frac{\partial(r\rho_A v_{Ar})}{\partial r} - \frac{1}{r}\frac{\partial(\rho_A v_{A\theta})}{\partial \theta} - \frac{\partial(\rho_A v_{AZ})}{\partial z} + r_A \quad (\text{cylindrical coordinates})$$

$$\frac{\partial \rho_A}{\partial t} = -\frac{1}{r^2}\frac{\partial(r^2\rho_A v_{Ar})}{\partial r} - \frac{1}{r\sin\theta}\frac{\partial(\sin\theta \rho_A v_{A\theta})}{\partial \theta} - \frac{1}{r\sin\theta}\frac{\partial(\rho_A v_{A\varphi})}{\partial \varphi} + r_A \quad (\text{spherical coordinates})$$

Compared to the CCS expression, the more complex form of the cylindrical and spherical coordinate expressions for the divergence term (the convection term) results from the curvilinear nature of these coordinates; i.e., the angular coordinate variables (such as θ and ϕ) change along curves, not lines.

The differential equation of continuity (total mass balance) derived in fluid mechanics for single component systems also applies to multicomponent systems in which chemical reactions happen.

To prove this is straightforward, and begins by summing equation ($\frac{\partial \rho_A}{\partial t} + \nabla \cdot \rho_A \mathrm{v}_A + r_A$) over all species present in solution,

$$\sum_{i=1}^{n}\frac{\partial \rho_i}{\partial t}=-\nabla\cdot\sum_{i=1}^{n}\rho_i\,\mathrm{v}_i+\sum_{i=1}^{n}r_i$$

Interchanging the summation and ∂/∂t operations on the left hand side and making use of the relation:

$$\sum_{i=1}^{n}\rho_i=\rho$$

the left hand side of equation ($\sum_{i=1}^{n}\frac{\partial \rho_i}{\partial t}=-\nabla\cdot\sum_{i=1}^{n}\rho_i\,\mathrm{v}_i+\sum_{i=1}^{n}r_i$) becomes:

$$\frac{\partial\sum_{i=1}^{n}p_i}{\partial t}=\frac{\partial\rho}{\partial t}$$

Furthermore, using the definition of the mass average velocity v,

$$\mathrm{v}=\sum_{i=1}^{n}\omega_i\mathrm{v}_i$$

and the fact that the mass fraction ωi = ρi/ρ,

$$\rho\mathrm{v}=\rho\sum_{i=1}^{n}\omega_i\mathrm{v}_i\sum_{i=1}^{n}\rho_i\mathrm{v}_\mathrm{i}$$

Using equation ($\rho\mathrm{v}=\rho\sum_{i=1}^{n}\omega_i\mathrm{v}_i\sum_{i=1}^{n}\rho_i\mathrm{v}_\mathrm{i}$), the convection term $-\nabla\cdot\sum_{i=1}^{n}\rho_i\mathrm{v}_\mathrm{i}$ in equation ($\sum_{i=1}^{n}\frac{\partial \rho_i}{\partial t}=-\nabla\cdot\sum_{i=1}^{n}\rho_i\,\mathrm{v}_i+\sum_{i=1}^{n}r_i$) becomes:

$$-\nabla\cdot\sum_{i=1}^{n}\rho_i\mathrm{v}_\mathrm{i}=-\nabla\cdot\rho\mathrm{v}$$

Finally, the chemical reaction term in equation ($\nabla\cdot\sum_{i=1}^{n}\rho_i\mathrm{v}_\mathrm{i}$) is required to evaluate to zero:

$$\sum_{i=1}^{n}r_i=0$$

Since mass is not produced or destroyed in chemical reactions (nuclear reactions can interconvert mass and energy, but this case is not being considered). Substituting equations ($\sum_{i=1}^{n}r_i=0$), ($-\nabla\cdot\sum_{i=1}^{n}\rho_i\mathrm{v}_\mathrm{i}=-\nabla\cdot\rho\mathrm{v}$), and ($\frac{\partial\sum_{i=1}^{n}p_i}{\partial t}=\frac{\partial\rho}{\partial t}$) into equation ($\nabla\cdot\sum_{i=1}^{n}\rho_i\mathrm{v}_\mathrm{i}$) leads to:

$$\frac{\partial\rho}{\partial t}=-\nabla\cdot\rho\mathrm{v}$$

Equation ($\frac{\partial \rho}{\partial t} = -\nabla \cdot \rho v$) is the differential equation of continuity familiar from fluid mechanics. This equation states the law of mass conservation; even in multicomponent systems, even if chemical reactions are present, the total accumulation of mass at a point (left hand side) can only occur by convection of mass to that point (right hand side). For multicomponent systems whose density ρ is constant (i.e. ρ does not vary from point to point irrespective of variations that may be present in temperature, pressure, or composition), equation ($\frac{\partial \rho}{\partial t} = -\nabla \cdot \rho v$) again simplifies to the condition for incompressible systems,

$$\nabla \cdot v = 0$$

Diffusion

Diffusion is a mass transfer phenomenon that causes the distribution of a chemical species to become more uniform in space as time passes.

In this case, species is a chemical dissolved in a solvent or a component in a gas mixture, such as the oxygen in air. The mass transfer of a species is the evolution of its concentration in space and time. If the concentration of a species is initially not uniform (the concentration might be greater in one region of a vessel than another, for example) then, over time, diffusion causes mass transfer in favor of a more uniform concentration.

The driving force for diffusion is the thermal motion of molecules. At temperatures above absolute zero, molecules are never at rest. Their kinetic energy means that they are always in motion, and when molecules collide with each other frequently, the direction of the motion becomes randomized. In most cases, these collisions are common; even in air at atmospheric pressure, which hardly seems a "dense" fluid, each molecule collides with a neighbor every few nanoseconds.

Three types of diffusion are distinguished, viz., molecular, *Brownian*, and turbulent. *Molecular diffusion* occurs in gases, liquids, and solids; both diffusion of molecules of extraneous substances (impurities) and self-diffusion are observed. Molecular diffusion occurs as a result of thermal motion of the molecules. It proceeds at a maximum rate in gases, at a lower rate in liquids, and at a still lower rate in solids—these differences being accounted for by the nature of thermal motion in these media.

In a gaseous phase, molecules possess a certain mean velocity depending on the temperature, but their motion is chaotic and in colliding, they change the direction of this motion. However, on the whole, the molecules of the substance migrate at a velocity much lower than the mean velocity of the molecular free motion. The higher the pressure, the denser is the molecule packing, the less is the free-path length, and the slower is the diffusion. The same occurs as molecule mass and size increase. Conversely, elevation of temperature causes an increase in the free-path length, a decrease in the number of collisions, and growth of free-motion velocity. These factors all lead to a speed-up of diffusion.

In liquids, molecular diffusion occurs by jumps of the molecules from one position to another; this arises when the energy of the molecule is high enough to rupture the bonds with the neighboring molecules allowing the molecule to move. On average, the jump does not exceed an intermolecular spacing, and since in a liquid this is much less than in a gas, the diffusion is substantially lower. Since a liquid is virtually incompressible, the diffusion rate is independent of pressure. Elevation of temperature increases intermolecular spacings and the velocity of vibrations and jumps of molecules, which enhances diffusion.

Gases contained in solids diffuse as ions or atoms migrating through interstitials of the crystal lattice, and the same is observed for atoms and ions with a radius much smaller than that of the ion or atom of the base substance constituting the solid. Diffusion of solid impurities occurs by interchange of sites of atoms and vacancies (unoccupied sites of crystal lattice), by migration of atoms through interstitials, by a simultaneous cyclic migration of several atoms, by a direct interchange of sites of two neighboring atoms, etc. Each displacement requires imparting to a particle a definite amount of energy (activation energy). Therefore, diffusion is extremely sensitive to temperature elevation, which manifests itself in its exponential dependence on temperature. Nevertheless, even at high temperature, diffusion in solids is much slower than in liquids.

Molecular Diffusion

In the general case, at a constant temperature and pressure, the molar diffusion flux $\dot{n}$ of substance A is proportional to the molar concentration gradient dC_A/dy and a one-dimensional formulation is described by an equation called Fick's law:

$$\dot{n}_A = -D_A \frac{dC_A}{dy},$$

where D_A is the diffusion coefficient. In a binary system, the flow of one component must be balanced by the counterflow of the other component:

$$nB = -D \frac{dC}{dy},$$

but since $C_A + C_B = const$, then,

$$\left|\frac{dC_A}{dy}\right| = \left|\frac{dC_B}{dy}\right| \text{ and } D_A = D_B = D_{AB}.$$

D_{AB} is said to be the interdiffusion coefficient.

In order to describe a unidirectional diffusion of A molecules in a multicomponent mixture of ideal gases, the Stefan-Maxwell equation:

$$\frac{dy_A}{dy} = \sum_{j=A}^{n} \frac{d_A C_j}{C_T^2 D_{A_j}} \left(u_j - u_A\right),$$

based on the kinetic theory of gases is used, where Y_A is the mole fraction of component A, $C_T = p/$

RT, the total concentration (density) of mixture; $C_A = pY_A/RT$, $c_j = pY_j/RT$, D_{Aj}, the interdiffusion coefficient for a pair A, j; and u_j and u_A are the diffusion rates for the respective components of the pair.

Brownian Diffusion

If fine particles (no more than a few microns in size) are placed in a stationary gas or liquid at rest, they randomly migrate in the bulk, the motion not decaying and being independent of the medium chemical properties. This phenomenon has come to be known as Brownian motion. It is brought about by the absence of a strict compensation of momentum in the molecule-particle collisions, i.e., the pressure pulsation affecting the particle. The suspended particles migrate independent of each other along intricate zigzag trajectories, changing direction up to 10^{14} times a second.

According to the general principles of statistical mechanics, the mean square value of displacement projection $\ddot{A}\overline{x}^2$ of a particle on an arbitrarily-chosen axis is proportional to the observation time t:

$$\Delta\overline{x}^2 = 2D_B t$$

where D_B is the diffusion coefficient of the Brownian particle.

For spherical particles of radius r, D_B is determined by the Stokes-Einstein equation:

$$D_B = \frac{kT}{6\pi\eta r},$$

where k is the Boltzmann constant; T, the temperature; K, η, the medium viscosity.

Equation ($\Delta\overline{x}^2 = 2D_B t$) has been obtained under the assumption that particle displacements in any direction are equiprobable and, consequently, the mean value of the products of particle displacement in nonoverlapping time intervals t_1 and t_2 is zero: $\overline{\Delta xt_1\, \Delta xt_2} = 0$. It holds true if we neglect the particle inertia, which is allowed for high enough t's.

If gas or liquid contains in its volume not a single particle but a great number of them, then, by virtue of the statistical character of the process, ultimately the particles appear to be uniformly distributed throughout the volume.

Turbulent Diffusion

In turbulent flows or in artificially-stirred media (e.g., by stirrers or packings), mass transfer occurs on a macro level when the finite gas or liquid volumes are displaced. However, under these conditions, mass transfer may also be proportional to the concentration gradient. This fact is indicative of turbulent (eddy) diffusion.

In a turbulent flow, eddies continuously form, break up, disappear and appear once again. Large-scale eddies may include small-scale ones. The element of substance transferred is not a single molecule of the substance but some quantity of it that depends on the eddy size. The intensity of a turbulent transfer is considerably higher than that resulting from molecular motions. Treating a nonstationary diffusion of the substance from a point source (point o) yields a simple relation:

$$\overline{y}^2 = 2D_E t,$$

where $\overline{y}^2$ is the mean square value of particle displacement from the longitudinal axis running through point o; D_E, the eddy diffusion coefficient; and t, the time. It has been proven that D_E depends on the time of motion for the particle volume considered.

The total mass transfer (e.g., in mixing) is a result of the joint effect of molecular and eddy diffusion. It is assumed in calculations that both processes are additive, i.e., the coefficients of eddy D_E and molecular D_M diffusion are added: $D = D_E + D_M$.

Diffusion in Electrolyte Solutions

A substance in electrolyte solutions, particularly dilute ones, exists in the form of ions (cations and anions). The theory of salt diffusion in dilute aqueous solutions is quite well-established, and the diffusion coefficient under an inifinite dilution is determined by the Nernst-Heckell equation:

$$D^0_{AB} = \frac{RT}{Fa^2}\frac{1/n_+ + 1/n_-}{1/\lambda^0_+ + 1/\lambda^0_-},$$

where D^0_{AB} is the diffusion coefficient; (m²/s); T, the temperature, K; Fa, the Faraday number; n_+ and n_-, the cation and anion valencies; λ^0_+ and λ^0_-, the limiting ionic conduction of the cation and the anion at a given temperature, m²/ohmЧmol.

With increasing solution concentration, D^0_{AB} first decreases and then grows.

In mixed electrolyte systems with a light cation, e.g. H^+, the latter may migrate leaving behind the anion, and the absence of current is due to another heavier, cation lagging. In these systems, a unidirectional diffusion proceeds due to a joint action of electrical and concentration gradients:

$$\dot{m}_+ = \frac{\lambda^0_+}{Fa^2}\left[-RT\frac{dc_+}{dy} Fa\, c_+ \frac{dE}{dy}\right],$$

$$\dot{m}_+ = \frac{\lambda^0_-}{Fa^2}\left[-RT\frac{dc_-}{dy} Fa\, c_- \frac{dE}{dy}\right],$$

$$\sum \dot{m}_+ = \sum \dot{m}_-, \sum \dot{m} = 0.$$

In a multiion system, it is important to allow for ion interaction when several ionic conductions greatly differ from other ionic conductions, e.g., H^+ and OH^-, for H^+ λ^0_+ = 349.8 cm²/WΩmol, for OH^- λ^0_- = 197.6 cm²/Ω mol, while for the rest ions of multiion systems ionic conductions are under 100.

Diffusion in Porous Bodies

There exist three mechanisms, viz., molecular or bulk diffusion, Knudsen diffusion, and surface diffusion. These can all come into action simultaneously in the same system.

Molecular diffusion is predominant in solids with large pores, whose size is much more than the free-path of the diffusing gas molecules.

Knudsen diffusion occurs in gas-filled solids with small pores, or under low pressure when the mean free-path of molecules is more than the pore size and the molecules collide with the walls more often than between themselves. Molecule reflection from the walls is normally diffuse, i.e., the molecules recoil in all directions, and diffusion of molecules along pores depends precisely on these collisions. The roles Knudsen and molecular diffusions perform are commensurable within a certain range of pore sizes and gas pressures.

Surface diffusion is observed during adsorption of a diffusing substance by a solid. Since the equilibrium surface gas concentration increases with an increase in partial pressure of the adsorbed species, a surface concentration gradient of a diffusing substance appears in the surface layer of a pore. Under certain conditions, this may enhance the total flow of a diffusing component.

For molecular diffusion in porous solids, use is made of the effective diffusion coefficient, arbitrarily related to the concentration gradient of a substance diffusing normal to the external surface of the body:

$$D_{eff}^{M} = D_{M}\,\theta/\xi$$

where D_M is the molecular diffusion coefficient, θ is a free (open) porosity and ξ, the so-called *tortuosity factor* determined in general experimentally in each specific case.

In order to calculate the flux density due to the Knudsen diffusion, it is also advisable to use some effective coefficient D_{eff}^{k}:

$$D_{eff}^{k} = \frac{8\theta^2}{3\xi S_{g\rho s}}\left(\frac{2RT}{\pi\rho}\right)^{0.5} = 19.400\,\frac{\theta^2}{\xi S_{g\rho s}}\left(\frac{T}{\overline{\mu}}\right)^{0.5},$$

where θ and ξ are the values of open porosity and the tortuosity factor; ρ_s and S_g, the density of a porous solid and its specific surface; and $\overline{\mu}$, the molecular mass of a diffusing substance.

In a transition region in which both molecular and Knudsen diffusions are essential, an overall effective diffusion coefficient is recommended:

$$D_{eff} = \left(\frac{1}{D_{eff}^{M}} + \frac{1}{D_{eff}^{k}}\right)^{-1}$$

The above refers only to a single-component diffusion. Multicomponent diffusion in porous solids has been inadequately investigated and data of any reliability for its calculation are not available. In considering the role of surface diffusion, it is generally assumed that the mass flux is proportional to the concentration gradient, and the absorbed layer of this sustance is extremely thin and does not change the pore cross section. Under these assumptions, the mass flux is:

$$\dot{m} = -\left(\frac{1}{D_{eff}^{M}} + \frac{1}{D_{eff}^{k}}\right)^{-1} \frac{dc}{dy} - D_{sp} \frac{d\left(s_g \rho_a c_{sur}\right)}{dy}$$

where D_{sp} is the surface diffusion coefficient and c_{sur} the surface concentration of adsorbed substance (the product $S_g \rho_s c_{sur}$ is the quantity of the adsorbed substance in a unit volume of porous mass). Assuming that an adsorbed layer and a gaseous phase in pores are at equilibrium and that the adsorption isotherm is linear, i.e., $S_g \rho_s c_{sur} = K_c$, then,

$$\dot{m} = -\left[\left(\frac{1}{D_{eff}^{M}} + \frac{1}{D_{eff}^{k}}\right)^{-1} + DK_{sp}\right] \frac{dc}{dy}.$$

This equation describes the diffusion flux in a porous solid when all the three diffusion mechanisms come into action simultaneously.

Mass Transfer Coefficients

Mass transfer coefficient is just analogous to the heat transfer coefficient. It is a proportionality constant to the difference in the concentrations and the rate of mass transfer. It determines the rate of mass transfer across a medium in response to a concentration gradient. It is an important principle in fluid dynamics and chemical engineering, especially in the design of fluid contact equipment such as gas scrubbers.

Mass transfer coefficients are usually obtained by semi-empirical methods. These combine a theoretical analysis of the problem using mathematical analogies and frequently guesses based on dimensional analysis, in combination with experimental data.

There are two main mechanisms depending on whether the flow is laminar or turbulent. In laminar flow (or in the absence of flow) the process of mass transfer is mainly due to molecular diffusion. If turbulent flow is involved the turbulent eddies act as a transport mechanism giving rise to transport rates that may be orders of magnitude higher than those due to molecular effects.

Overall Mass Transfer Coefficients

Experimentally the mass transfer film coefficients k_y and k_x are difficult to measure except for cases where the concentration difference across one phase is small and can be neglected. Under these circumstances, the overall mass transfer coefficients K_y and K_x are measured on the basis of the gas phase or the liquid phase. The entire two-phase mass transfer effect can then be measured in terms of gas phase molar fraction driving force as:

$$N_A = K_y\left(y_{AG} - y_A^*\right)$$

where, K_y is based on the overall driving force for the gas phase, in mole/m² .s and y_A^* is the value of concentration in the gas phase that would be in the equilibrium with x_{AL}. Similarly, the entire two-phase mass transfer effect can then be measured in terms of liquid phase molar fraction driving force as:

$$N_A = K_x\left(x_A^* - x_{AL}\right)$$

where K_x is based on the overall driving force for the liquid phase, in mole/m² .s and x_A^* is the value of concentration in the liquid phase that would be in the equilibrium with y_{AG}. A relation between the overall coefficients and the individual mass transfer film coefficients can be obtained when the equilibrium relation is linear as $y_{Ai} = mx_{Ai}$. The linear equilibrium condition can be obtained at the low concentrations, where Henry's law is applicable. Here the proportionality constant m is defined as m= H/P. Utilizing the relationship, $y_{Ai} = mx_{Ai}$, gas and liquid phase concentrations can be related by:

$$y_A^* = mx_{AL}$$

and

$$y_{AG} = mx_A^*$$

Rearranging equation $N_A = K_y\left(y_{AG} - y_A^*\right)$, one can get:

$$\frac{1}{K_y} = \frac{y_{AG} - y_A^*}{N_A}$$

From geometry, $y_{AG} - y_A^*$ can be written as:

$$y_{AG} - y_A^* = \left(y_{AG} - y_{Ai}\right) + \left(y_{Ai} - y_A^*\right)$$

Substituting equation $y_{AG} - y_A^* = \left(y_{AG} - y_{Ai}\right) + \left(y_{Ai} - y_A^*\right)$ in equation $\frac{1}{K_y} = \frac{y_{AG} - y_A^*}{N_A}$

$$\frac{1}{K_y} = \frac{y_{AG} - y_A^*}{N_A} = \frac{\left(y_{AG} - y_{Ai}\right)}{N_A} + \frac{\left(y_{Ai} - y_A^*\right)}{N_A} = \frac{\left(y_{AG} - y_{Ai}\right)}{N_A} + \frac{m\left(x_{Ai} - x_{AL}\right)}{N_A}$$

The substitution of equation $N_A = K_y\left(y_{AG} - y_A^*\right)$ into the equation:

$$\frac{1}{K_y} = \frac{y_{AG} - y_A^*}{N_A} = \frac{\left(y_{AG} - y_{Ai}\right)}{N_A} + \frac{\left(y_{Ai} - y_A^*\right)}{N_A} = \frac{\left(y_{AG} - y_{Ai}\right)}{N_A} + \frac{m\left(x_{Ai} - x_{AL}\right)}{N_A}$$

relates overall gas phase mass transfer coefficient (K_y) to the individual film coefficients by,

$$\frac{1}{K_y} = \frac{1}{k_y} + \frac{m}{k_x}$$

Similarly the relation of overall liquid phase mass transfer coefficient (K_x) to the individual film coefficients can be derived as follows:

$$\frac{1}{K_x} = \frac{x_A^* - x_{AL}}{N_A} = \frac{y_{AG} - y_{Ai}}{mN_A} + \frac{x_{Ai} - x_{AL}}{N_A}$$

Or

$$\frac{1}{K_x} = \frac{1}{mk_y} + \frac{1}{k_x}$$

The following relationships between the mass transfer resistances can be made from the equations $\frac{1}{K_y} = \frac{1}{k_y} + \frac{m}{k_x}$ and $\frac{1}{K_x} = \frac{1}{mk_y} + \frac{1}{k_x}$:

$$\frac{\text{Resistance in gas phase}}{\text{Total resistance in both phases}} = \frac{1/k_y}{1/K_y}$$

$$\frac{\text{Resistance in liquid phase}}{\text{Total resistance in both phases}} = \frac{1/k_x}{1/K_x}$$

If solute A is very soluble in the liquid, m is very small. Then the term m/k_x in equation $\frac{1}{\ddot{u}} = \frac{1}{\ddot{u}} + \frac{m}{}$ becomes minor and consequently the major resistance is represented by $1/k_y$. In this case, it is said that the rate of mass transfer is gas phase controlled. In the extreme it becomes:

$$\frac{1}{K_y} \approx \frac{1}{k_y}$$

The total resistance equals the gas film resistance. The absorption of a very soluble gas, such as ammonia in water is an example of this kind. Conversely when solute A is relatively insoluble in the liquid, m is very large. Consequently the first term of equation $\frac{1}{K_x} = \frac{1}{mk_y} + \frac{1}{k_x}$ becomes minor and the major resistance to the mass transfer resides within the liquid. The system becomes liquid film controlling. Finally this becomes:

$$\frac{1}{K_x} \approx \frac{1}{k_x}$$

The total resistance equals the liquid film resistance. The absorption of a gas of low solubility, such as carbon dioxide or oxygen in water is of this type of system.

Example: In an experimental study of the absorption of ammonia by water in a wetted-wall column, the value of overall mass transfer coefficient, KG was found to be 2.75×10^{-6} kmol/m² -s-kPa. At one point in the column, the composition of the gas and liquid phases were 8.0 and 0.115 mole% NH^3, respectively. The temperature was 300K and the total pressure was 1 atm. Eighty five % of the total resistance to mass transfer was found to be in the gas phase. At 300 K, Ammonia –water solutions follows Henry's law upto 5 mole% ammonia in the liquid, with m = 1.64 when the total pressure is 1 atm. Calculate the individual film coefficients and the interfacial concentrations. Interfacial concentrations lie on the equilibrium line.

Solution: The first step in the solution is to convert the given overall coefficient from K_G to K_y.

$$K_y = K_G\, P = 2.75 \times 10^{-6} \times 101.3 = 2.786 \times 10^{-4}\ \text{kmol/m}^2 - \text{s}$$

For a gas-phase resistance that accounts for 85% of the total resistance,

$$k_y = \frac{K_y}{0.85} = 3.28 \times 10^{-4}\ \text{kmol/m}^2 - s$$

From equation, $\frac{1}{K_y} = \frac{1}{k_y} + \frac{m}{k_x}$, by substituting the values of K_y, k_y and m $k_x = 3.05 \times 10^{-3}$ kmol/m² -s

To estimate the ammonia flux and the interfacial concentrations at this particular point in the column use the equation, $y_A^* = mx_{A,L}$ to calculate:

$$y_A^* = mx_{A,L} = 1.64 \times 1.15 \times 10^{-3} = 1.886 \times 10^{-3}$$

The flux is from equation:

$$N_A = K_y \left(y_{AG} - y_A^* \right) = 2.768 \times 10^{-4} \left(0.080 - 1.866 \times 10^{-3} \right) = 2.18 \times 10^{-5}\ kmol / m^2 - s$$

Calculate the gas-phase interfacial concentration from equation, $N_A = k_y \left(y_{AG} - y_{A,i} \right)$ as:

$$y_{A,i} = y_{AG} - \frac{N_A}{k_y} = 0.080 - \frac{2.18 \times 10^{-5}}{3.28 \times 10^{-4}} = 0.01362$$

Since the interfacial concentrations lie on the equilibrium line,

$$x_{A,i} = \frac{y_{A,i}}{m} = \frac{0.01362}{1.64} = 8.305 \times 10^{-3}$$

References

- Heat-transfer-convection, physics: byjus.com, Retrieved 29 June, 2019
- Mass-transfer-coefficient, properties: calculator.org, Retrieved 14 July, 2019
- What-is-diffusion, multiphysics: comsol.co.in, Retrieved 11 January, 2019
- Convective-heat-transfer-d: engineeringtoolbox.com, Retrieved 18 April, 2019

5

Applications of Chemical Engineering

Chemical engineering is applied in numerous fields such as agriculture and food industry. Within agriculture, it is used to produce agricultural chemicals like insecticides, fungicides and pesticides. The diverse applications of chemical engineering in these sectors have been thoroughly discussed in this chapter.

Application of Chemical Engineering in Food Industry

In the food industry, there are various principles which use the phenomenon of chemical engineering. The chemical engineers contribute to the food industry by fulfilling our demands and inventing techniques that enhance the quality and increase the shelf life of the food. They also come up with technologies that can protect us against food-borne diseases.

- Enhance fruit and vegetable production: The chemical engineers contribute to the food industry by finding ways and inventing pesticides and fertilizers that protect and help to enhance food production.
- Food packaging: Chemical engineers have a significant role to play in packaging. They develop advanced materials and techniques that can be used for packaging. Packaging is one of the most critical trends in the food industry, and chemical engineers are responsible for it. They ensure that packaging material is free from artificial additives. In recent years, they have made developments which have reduced the threat of food packaging and have increased the shelf life of the packed food.
- Protecting Perishable foods: We can credit chemical engineers for handling and safeguarding perishable foods. They come up with new sterilization techniques that can help preserve food against spoilage and help in maintaining the quality of the food by minimizing the risk of the deterioration.
- Innovation in food: The chemical engineers formulate new products by mixing and processing different industries. They modify the ingredients in the foods we eat and change the ingredients for the better flavors. They find alternative elements. All-in-all we can say that chemical engineers are committed to the betterment of human nutrition. They apply chemistry and engineering together and come up with food which is more fresh, tasty, wholesome, and ready to eat.
- Safe food handling practices: Food safety is incredibly important as it helps in preserving the food against food-borne diseases. The chemical engineers are responsible for inventing techniques that can help in handling process and can help us increase the safety of the food.

Chemicals used in Food Processing

Chemical food additives preserve shelf life by reducing or eliminating the growth of microorganisms that cause food decay. These added chemicals are controversial as some may have a detrimental effect on your health. Eating more fresh, whole foods may be a safe alternative to avoiding these chemical additives used in food processing.

Nitrites and Nitrates

Nitrites and nitrates are effective antimicrobials that protect against the botulism bacterium; however, they are known to be a health hazard. According to Richard Scanlan, of the Linus Pauling Institute, nitrites react with certain amino acids in the digestive system causing the formation of nitrosamines, substances known to cause cancer. Sodium nitrate, used as a color stabilizer, slowly metabolizes into nitrites. Certain brands of products commonly containing nitrites and nitrates are bacon, cured meats, tobacco products and hot dogs.

Sulfites and Sulphur Dioxide

Sulfites and sulphur dioxide are both antimicrobials, used primarily as preservatives. Sulfites are found predominately in dried fruits and red wine, and sometimes in certain brands of fruit juices. Sulphur dioxide is used in numerous foods. This compound slows the oxidation of fats called lipids. When sprayed on freshly harvested fruits and vegetables, it stops the natural ripening process. These two compounds can cause an allergic reaction in certain individuals and is a known allergen for many asthmatics.

Alginate and Propylene Glycol Alginate

Alginate is a naturally derived compound of kelp, or seaweed, and is used as a thickening agent and foam stabilizer. However, propylene glycol alginate is a chemically modified substance, utilizing algin to thicken foods with a higher acid content such as beer, soft drinks and salad dressing. Alginate is used in milk products such as ice cream, cheese, cottage cheese and yogurt.

Aspartame

This compound is one that should be avoided. Aspartame is a combination of amino acids and methanol. This chemical compound may cause cancer and other health-related problems, though more intensive studies are needed to confirm a connection. Used as an artificial sweetener, Aspartame is used in diet foods in place of sugar, and used primarily in soft drinks and drink mixes, and many low-calorie, processed fresh and frozen desserts.

Chemical Engineering in Daily Life

The products that we buy and use every day, and often take for granted, have been chemically engineered so that they fulfil their required function.

And as you know, the best way to explain something is by way of example.

The products are also a great way to introduce young people to a potential career in chemical engineering .

Hairspray

This hair care product is incorporated into many people's morning routines. Both men and women want to achieve a desired hairstyle. So it is important that hairspray is designed to provide a strong hold that lasts throughout the day.

Hairspray consists of complex polymers that have the ability to form invisible bonds across our hair's fibre intersection. The choice of these complex polymers in hairspray directly relates to consumer need, so vinyl acetate is used to ensure strong hold and, in addition, is resistant to humidity.

Hairspray is usually packaged in an aerosol can powered by a pressurised gas – dimethyl ether. This gas acts as a propellant within the aerosol and is boiled away when the nozzle is pushed down. The rest of the ingredients are then propelled out of the can in the form of a fine mist.

Hairspray is produced in a batch process in large tanks of around 9,000 L where the ingredients are mixed by large turbine blades – the main base of this product consists of water and a denatured alcohol.

The other ingredients that are added to perform specific functions include perfume oil (pretty self-explanatory), amino methyl propanol which controls the acidity of the product, cyclopentasiloxane which acts as a thickener, lubricant and water repellent to give hairspray its 'silky feel', and borate esters that prevent the aerosol can from rusting.

Toilet Paper

Every household will need its supply of toilet paper and so this makes the product a must-buy consumer product.

Toilet paper can be made from either 'virgin' or new paper (formed of chipped wood) or from recycled paper. The other materials used in its manufacture include water, chemicals for breaking down the raw materials into usable fibres and bleaches.

When the raw materials include recycled paper, the batch process starts by mixing all the different types of paper and then removing the ink by injecting the pulp with air – the ink particles can then be skimmed off as they rise to the top of the pulp mixture. If the feed into the process is 'virgin' or new paper, the chipped wood is cooked within a pressure cooker in order to evaporate some of the excess water and reduce the wood into pulp.

The pulp mixture, whether from 'virgin' or recycled paper having now been cleaned, is passed through some heated rollers to reduce the moisture content to approximately five per cent so that it is ready for bleaching. Before bleaching, the mixture is chopped back up into pulp.

Once bleached, the wet pulp passes through a high temperature dryer and is rolled into paper which is one ply thick. From this point, the paper can be embossed with a pattern which actually increases its thickness.

Automated machines then do the final job of unwinding, cutting and then rewinding the paper onto cardboard tubing. Rolls of toilet paper are then cut and wrapped into the final product that we see line the shelves of the supermarkets.

Deodorant

Almost everyone uses deodorant at different times throughout the day and usually they are delivered to the consumer as either roll-on deodorants, sticks or aerosol sprays, to be applied usually to the underarm area.

Deodorant is designed to prevent body odour, and antiperspirant deodorants are designed to prevent sweating – a subtle difference but the desired outcome is to keep the individual smelling fresh.

Deodorants are typically manufactured in a batch process within high shear mixers that have a volume capacity approaching 2,000 L – the formulation and combination of ingredients really depends on what form the product is going to be delivered to the consumer.

Solvents or silicones are mixed with the powder ingredients such as cetyl alcohol into the oil phase within the mixer. The active ingredient is prepared separately in an aqueous, dispersed phase.

Both the oil phase and dispersed phase are combined and this forms an emulsion. If the end product is a delivered as a stick or roll-on, then gelling agents or polymers are added. Otherwise, perfume/fragrance is the next ingredient before the product is cooled and then poured into containers to be packaged.

The active ingredients in deodorant differ from that of antiperspirants because of the different ways they inhibit odour. Aluminium based compounds are usually the active ingredients for antiperspirants, with aluminum zirconium tetrachlorohydrex glycine being the most commonly used.

Batch testing the products for safety and efficacy is an important part of the process, as products often land on our shelves with labels that state such things as '48 hour protection' or 'contains aloe vera'. Manufacturers need to make sure that their products are safe, functional and contain exactly what it says on the container.

Detergent/Washing Powder

Detergent or washing powder is a staple product that's found within most households in order to clean clothes, linen and other textiles. When the powder is dissolved in water i.e. within a washing machine, it aids the removal of dirt from a textile surface.

Detergent powder can be produced in either a batch or continuous process, with the larger manufacturers favouring the continuous process, also known as the agglomeration process.

Detergents are based on the surface active ingredient or surfactant. Surfactants consist of hydrocarbon chain and a hydrophilic ionic or non-iconic group, both of which perform a specific function.

The hydrocarbon chain part of the surfactant molecule attaches itself to dirt and the hydrophilic end attaches itself to water. This means that the dirt is chemically removed by the hydrocarbon chain into the water. The most common surfactants used in industry are anionic and usually come in the form of water-soluble sodium salts.

Enzymes and perfumes are other ingredients of detergent. The enzymes are present to break down stains and the perfumes are added to ensure that the laundry smells nice, and covers the chemical smell of the detergent.

The agglomeration process starts with the dry ingredients being fed into an agglomerator where blades mix the material into a fine consistency. Liquid ingredients are then added to the mix via spray nozzles directly into the agglomerator. This reaction produces heat which leads to the formation of a hot, viscous liquid.

Once the liquid exits the agglomerator, hot air blowers are applied and the detergent collects on a drying belts. This is then pulverized and pushed through sizing screens (to avoid large lumps of powder) to form the detergent/washing powder product that then goes on to be packaged. If the detergent is a liquid product, the powder is simply blended with a solution of water and solubilizers – the solubilizers aid the mixing of the detergent and water.

Toothpaste

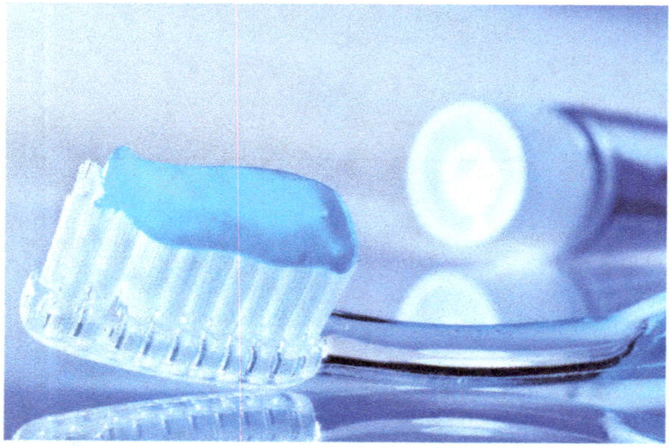

Dentists and dental hygienists recommend that we brush our teeth twice a day for two minutes using a fluoride toothpaste to prevent gum disease, decay and build up of plaque.

Toothpaste has become an essential product, often complimented with the use mouthwash and dental floss. An important ingredient of toothpaste is fluoride. Fluoride acts as the active ingredient – it prevents decay and the formation of cavities by increasing the strength of teeth. Sodium fluoride is the most common fluoride compound used in the manufacture of toothpaste. Other important ingredients of toothpaste include: binders, which acts as a thickener; abrasives, which loosen the plaque from the teeth; foaming agents (surfactants); which, which retain water; flavours; colourants and sweeteners.

Toothpaste manufacture is a batch process, where ingredients are weighed before mixing. Each batch can produce an equivalent of 10,000 tubes of toothpaste.

An important feature of some toothpastes are the different coloured stripes. Artificial dyes are added to a batch to change its colour. Essential oils are then used to add these colours (which often attribute to a different flavour) to the product at the packaging stage.

The final product is filled into tubes, with the use of automated equipment and pumps and it is at this stage that the product is tested for safety and efficacy.

Application of Chemical Engineering in Agriculture

Chemical engineering has played a leading role in improving production in the field of agriculture. Chemical fertilizers which paved the way for green revolution in India are a contribution from chemical engineering. It has made India self sufficient in terms of producing cereals and grains which forms the food crops.

Agricultural Chemicals

Agricultural chemicals are defined as fungicides and insecticides that are used to control crop-harming organisms (e.g., fungi, nematodes, mites, insects, and rodents) or viruses (hereinafter collectively referred to as "diseases and pests") (the "crop," as used herein, shall include wood and agroforestry products and those used to promote or inhibit the physiology of agricultural and other products, such as plant growth regulators and germination inhibitors (Hereinafter referred to as "agricultural and other products). The chemical agents here include those manufactured using the above chemical agents as raw materials or ingredients that are intended to control the diseases and pests that are stipulated in the ordinances. The law also includes "natural enemies" and "microorganisms" that are used to control diseases and pests of agricultural and other products, as the agricultural chemicals.

Table: Agricultural chemicals are classified as follows by application target.

A Insecticides	Agents for controlling harmful insect pests that damage field crops.
B Fungicides	Agents for controlling diseases that damage field crops.
C Insect-fungicides	Agents that simultaneously control harmful insect pests and diseases that damage field crops.
D Herbicides	Agents for controlling weeds.
E Rodenticides	Agents for controlling rats and other rodents.
F Plant growth regulators	Agents to promote or inhibit the growth of field crops.
G Attractants	Agents that attract mainly harmful insect pests by odor or other means.
H Repellents	Agents for having repellent action on harmful mammals and birds that damage field crops.
I Spreaders	Agents that are mixed with other agricultural chemicals to enhance the adherence of these chemicals.

Insecticides

Insecticide is any toxic substance that is used to kill insects. Such substances are used primarily to control pests that infest cultivated plants or to eliminate disease-carrying insects in specific areas.

Insecticides can be classified in any of several ways, on the basis of their chemistry, their toxicological action, or their mode of penetration. In the latter scheme, they are classified according to whether they take effect upon ingestion (stomach poisons), inhalation (fumigants), or upon penetration of the body covering (contact poisons). Most synthetic insecticides penetrate by all three of these pathways, however, and hence are better distinguished from each other by their basic chemistry. Besides the synthetics, some organic compounds occurring naturally in plants are useful insecticides, as are some inorganic compounds; some of these are permitted in organic farming applications. Most insecticides are sprayed or dusted onto plants and other surfaces traversed or fed upon by insects.

Modes of Penetration

Stomach poisons are toxic only if ingested through the mouth and are most useful against those insects that have biting or chewing mouth parts, such as caterpillars, beetles, and grasshoppers. The chief stomach poisons are the arsenicals—e.g., Paris green (copper acetoarsenite), lead arsenate, and calcium arsenate; and the fluorine compounds, among them sodium fluoride and cryolite. They are applied as sprays or dusts onto the leaves and stems of plants eaten by the target insects. Stomach poisons have gradually been replaced by synthetic insecticides, which are less dangerous to humans and other mammals.

Contact poisons penetrate the skin of the pest and are used against those arthropods, such as aphids, that pierce the surface of a plant and suck out the juices. The contact insecticides can be divided into two main groups: naturally occurring compounds and synthetic organic ones. The naturally occurring contact insecticides include nicotine, developed from tobacco; pyrethrum, obtained from flowers of Chrysanthemum cinerariaefolium and Tanacetum coccineum; rotenone, from the roots of *Derris* species and related plants; and oils, from petroleum. Though these compounds were originally derived mainly from plant extracts, the toxic agents of some of them (e.g., pyrethrins) have been synthesized. Natural insecticides are usually short-lived on plants and cannot provide protection against prolonged invasions. Except for pyrethrum, they have largely been replaced by newer synthetic organic insecticides.

Fumigants are toxic compounds that enter the respiratory system of the insect through its spiracles, or breathing openings. They include such chemicals as hydrogen cyanide, naphthalene, nicotine, and methyl bromide and are used mainly for killing insect pests of stored products or for fumigating nursery stock.

Synthetic Insecticides

The synthetic contact insecticides are now the primary agents of insect control. In general they penetrate insects readily and are toxic to a wide range of species. The main synthetic groups are the chlorinated hydrocarbons, organic phosphates (organophosphates), and carbamates.

Chlorinated Hydrocarbons

The chlorinated hydrocarbons were developed beginning in the 1940s after the discovery of the insecticidal properties of DDT. Other examples of this series are BHC, lindane, Chlorobenzilate, methoxychlor, and the cyclodienes (which include aldrin, dieldrin, chlordane, heptachlor, and endrin). Some of these compounds are quite stable and have a long residual action; they are, therefore, particularly valuable where protection is required for long periods. Their toxic action is not fully understood, but they are known to disrupt the nervous system. A number of these insecticides have been banned for their deleterious effects on the environment.

Organophosphates

The organophosphates are now the largest and most versatile class of insecticides. Two widely used compounds in this class are parathion and malathion; others are Diazinon, naled, methyl parathion, and dichlorvos. They are especially effective against sucking insects such as aphids and mites, which feed on plant juices. The chemicals' absorption into the plant is achieved either by spraying the leaves or by applying solutions impregnated with the chemicals to the soil, so that intake occurs through the roots. The organophosphates usually have little residual action and are important, therefore, where residual tolerances limit the choice of insecticides. They are generally much more toxic than the chlorinated hydrocarbons. Organophosphates kill insects by inhibiting the enzyme cholinesterase, which is essential in the functioning of the nervous system.

Carbamates

The carbamates are a group of insecticides that includes such compounds as carbamyl, methomyl, and carbofuran. They are rapidly detoxified and eliminated from animal tissues. Their toxicity is thought to arise from a mechanism somewhat similar to that for the organophosphates.

Fungicides

A fungicide is a substance, often a chemical, that is used to kill fungi, fungal spores, and fungal infections. There are both natural fungicides and synthetic fungicides available. In general, commercial fungicides, which are usually synthetic, are commonly used in horticulture in order to kill fungal pathogens that have infected plants.

Fungal pathogens are the No. 1 cause of crop loss around the world and can cause serious damage in agriculture, resulting in critical losses of yield, quality, and profit in both plants and animals.

In organic gardening, fungi are often treated by washing plants with a light, soapy solution, but this gentle approach is not practical for commercial agriculture. For this reason, commercial growers almost always choose to use a fungicide. Fungicides can either be contact, translaminar, or systemic in nature.

Contact fungicides are substances that are applied directly to infected leaves or stems. They only protect the area where the spray is deposited.

Translaminar fungicides are redistributed from the upper, sprayed leaf surfaces to the lower, unsprayed surfaces.

Systemic fungicides, which are drawn up through the roots and redistributed through the xylem vessels, move to all parts of a plant.

Pesticides

Pesticides are chemical substances that are meant to kill pests. In general, a pesticide is a chemical or a biological agent such as a virus, bacterium, antimicrobial, or disinfectant that deters, incapacitates, kills, pests.

This use of pesticides is so common that the term pesticide is often treated as synonymous with plant protection product. It is commonly used to eliminate or control a variety of agricultural pests that can damage crops and livestock and reduce farm productivity. The most commonly applied pesticides are insecticides to kill insects, herbicides to kill weeds, rodenticides to kill rodents, and fungicides to control fungi, mold, and mildew.

Pesticides are not recent inventions! Many ancient civilizations used pesticides to protect their crops from insects and pests. Ancient Sumerians used elemental sulfur to protect their crops from insects. Whereas, Medieval farmers experimented with chemicals using arsenic, lead on common crops.

The Chinese used arsenic and mercury compounds to control body lice and other pests. While, the Greeks and Romans used oil, ash, sulfur, and other materials to protect themselves, their livestock, and their crops from various pests.

Meanwhile, in the nineteenth century, researchers focused more on natural techniques involving compounds made with the roots of tropical vegetables and chrysanthemums. In 1939, Dichloro-Diphenyl-Trichloroethane (DDT) was discovered, which has become extremely effective and rapidly used as the insecticide in the world. However, twenty years later, due to biological effects and human safety, DDT has been banned in almost 86 countries.

The Food and Agriculture Organization (FAO) has defined pesticide as:

> "Any substance or mixture of substances intended for preventing, destroying, or controlling any pest, including vectors of human or animal disease, unwanted species of plants or animals, causing harm during or otherwise interfering with the production, processing, storage, transport, or marketing of food, agricultural commodities, wood and wood products or animal feedstuffs, or substances that may be administered to animals for the control of insects, arachnids, or other pests in or on their bodies."

Types of Pesticides

These are grouped according to the types of pests which they kill.

Grouped by types of pests they kill:

- Insecticides – insects
- Herbicides – plants
- Rodenticides – rodents (rats & mice)
- Bactericides – bacteria
- Fungicides – fungi
- Larvicides – larvae

Based on how biodegradable they are:

Pesticides can also be considered as:

- Biodegradable: The biodegradable kind is those which can be broken down by microbes and other living beings into harmless compounds.
- Persistent: While the persistent ones are those which may take months or years to break down.

Another way to classify these is to consider those that are chemical forms or are derived from a common source or production method.

Chemically-related Pesticides

- Organophosphate: Most organophosphates are insecticides, they affect the nervous system by disrupting the enzyme that regulates a neurotransmitter.
- Carbamate: Similar to the organophosphorus pesticides, the carbamate pesticides also affect the nervous system by disrupting an enzyme that regulates the neurotransmitter. However, the enzyme effects are usually reversible.
- Organochlorine insecticides: They were commonly used earlier, but now many countries have been removed Organochlorine insecticides from their market due to their health and environmental effects and their persistence (e.g., DDT, chlordane, and toxaphene).
- Pyrethroid: These are a synthetic version of pyrethrin, a naturally occurring pesticide, found in chrysanthemums(Flower). They were developed in such a way as to maximise their stability in the environment.
- Sulfonylurea herbicides: The sulfonylureas herbicides have been commercialized for weed control such as pyrithiobac-sodium, cyclosulfamuron, bispyribac-sodium, terbacil, sulfometuron-methyl Sulfosulfuron, rimsulfuron, pyrazosulfuron-ethyl, imazosulfuron, nicosulfuron, oxasulfuron, nicosulfuron, flazasulfuron, primisulfuron-methyl, halosulfuron-methyl,

flupyrsulfuron-methyl-sodium, ethoxysulfuron, chlorimuron-ethyl, bensulfuron-methyl, azimsulfuron, and amidosulfuron.

- Biopesticides: The biopesticides are certain types of pesticides derived from such natural materials as animals, plants, bacteria, and certain minerals.

References

- What-are-the-chemical-engineering-applications-in-food-industry, question: airtract.com, Retrieved 27 May, 2019
- List-of-chemicals-used-in-food-processing: livestrong.com, Retrieved 21 July, 2019
- Five-every-day-products-and-the-chemical-engineering-that-goes-into-them-day: ichemeblog.org, Retrieved 13 July, 2019
- Insecticide, technology: britannica.com, Retrieved 24 March , 2019
- Fungicide, definition: maximumyield.com, Retrieved 1 June, 2019
- Pesticides, chemistry: byjus.com, Retrieved 21 March , 2019

Permissions

All chapters in this book are published with permission under the Creative Commons Attribution Share Alike License or equivalent. Every chapter published in this book has been scrutinized by our experts. Their significance has been extensively debated. The topics covered herein carry significant information for a comprehensive understanding. They may even be implemented as practical applications or may be referred to as a beginning point for further studies.

We would like to thank the editorial team for lending their expertise to make the book truly unique. They have played a crucial role in the development of this book. Without their invaluable contributions this book wouldn't have been possible. They have made vital efforts to compile up to date information on the varied aspects of this subject to make this book a valuable addition to the collection of many professionals and students.

This book was conceptualized with the vision of imparting up-to-date and integrated information in this field. To ensure the same, a matchless editorial board was set up. Every individual on the board went through rigorous rounds of assessment to prove their worth. After which they invested a large part of their time researching and compiling the most relevant data for our readers.

The editorial board has been involved in producing this book since its inception. They have spent rigorous hours researching and exploring the diverse topics which have resulted in the successful publishing of this book. They have passed on their knowledge of decades through this book. To expedite this challenging task, the publisher supported the team at every step. A small team of assistant editors was also appointed to further simplify the editing procedure and attain best results for the readers.

Apart from the editorial board, the designing team has also invested a significant amount of their time in understanding the subject and creating the most relevant covers. They scrutinized every image to scout for the most suitable representation of the subject and create an appropriate cover for the book.

The publishing team has been an ardent support to the editorial, designing and production team. Their endless efforts to recruit the best for this project, has resulted in the accomplishment of this book. They are a veteran in the field of academics and their pool of knowledge is as vast as their experience in printing. Their expertise and guidance has proved useful at every step. Their uncompromising quality standards have made this book an exceptional effort. Their encouragement from time to time has been an inspiration for everyone.

The publisher and the editorial board hope that this book will prove to be a valuable piece of knowledge for students, practitioners and scholars across the globe.

Index

www.ingramcontent.com/pod-product-compliance
Lightning Source LLC
LaVergne TN
LVHW080653210926
848485LV00003B/369